Lecture Notes in Mathematics   2069

Editors:
J.-M. Morel, Cachan
B. Teissier, Paris

For further volumes:
http://www.springer.com/series/304

David Futer • Efstratia Kalfagianni
Jessica Purcell

Guts of Surfaces and the Colored Jones Polynomial

 Springer

David Futer
Department of Mathematics
Temple University
Philadelphia, PA 19122, USA

Efstratia Kalfagianni
Department of Mathematics
Michigan State University
East Lansing, MI 48824, USA

Jessica Purcell
Department of Mathematics
Brigham Young University
Provo, UT 84602, USA

ISBN 978-3-642-33301-9   ISBN 978-3-642-33302-6 (eBook)
DOI 10.1007/978-3-642-33302-6
Springer Heidelberg New York Dordrecht London

Lecture Notes in Mathematics ISSN print edition: 0075-8434
                         ISSN electronic edition: 1617-9692

Library of Congress Control Number: 2012953002

Mathematics Subject Classification (2010): 57N10, 57M25, 57M27, 57M50, 57M15, 57R56

Printed on acid-free paper

Springer is part of Springer Science+Business Media (www.springer.com)

Preface

Around 1980, W. Thurston proved that every knot complement satisfies the geometrization conjecture: it decomposes into pieces that admit locally homogeneous geometric structures. In addition, he proved that the complement of any non-torus, non-satellite knot admits a complete hyperbolic metric which, by the Mostow–Prasad rigidity theorem, is necessarily unique up to isometry. As a result, geometric information about a knot complement, such as its volume, gives topological invariants of the knot.

In the mid-1980s ideas from quantum physics led to powerful and subtle knot invariants, including the Jones polynomial and its relatives, the *colored* Jones polynomials. Topological quantum field theory predicts that these quantum invariants are very closely connected to geometric structures on knot complements and particularly to hyperbolic geometry. The *volume conjecture* of R. Kashaev, H. Murakami, and J. Murakami, which asserts that the volume of a hyperbolic knot is determined by certain asymptotics of colored Jones polynomials, fits into the context of these predictions. Despite compelling experimental evidence, these conjectures are currently verified for only a few examples of hyperbolic knots.

This monograph initiates a systematic study of relations between quantum and geometric knot invariants. Under mild diagrammatic hypotheses that arise naturally in the study of knot polynomial invariants (A- or B-adequacy), we derive direct and concrete relations between colored Jones polynomials and the topology of incompressible spanning surfaces in knot and link complements. We prove that the growth of the degree of the colored Jones polynomials is a boundary slope of an essential surface in the knot complement and that certain coefficients of the polynomial measure how far this surface is from being a fiber in the knot complement. In particular, the surface is a fiber if and only if a certain coefficient vanishes.

Our results also yield concrete relations between hyperbolic geometry and colored Jones polynomials: for certain families of links, coefficients of the polynomials determine the hyperbolic volume to within a factor of 4. In several instances, our methods provide a more intrinsic explanation for similar connections that have been previously observed.

The approach we take in this monograph is to generalize the checkerboard decompositions of alternating knots and links to links with A- or B-adequate diagrams. The analogues of the checkerboard surfaces in this generalized setting are the all-A or all-B state surfaces. For A- or B-adequate diagrams, we show that these state surfaces are incompressible and obtain an ideal polyhedral decomposition of their complement. This is done in Chaps. 2 and 3.

The main body of the monograph is Chaps. 4–6, where we study the Jaco–Shalen–Johannson (JSJ) decomposition of the state surface complement. Our results establish a dictionary between the pieces of the JSJ decomposition of the surface complement and the combinatorial structure of certain spines of the surface (state graphs). In particular, we give a combinatorial formula for the complexity of the hyperbolic part of the JSJ decomposition (the *guts*) of the surface complement in terms of the diagram of the knot and use this to give lower bounds on the volume of the knot complement. Since state graphs have previously appeared in the study of Jones polynomials, our setting and methods allow to derive relations between quantum invariants and geometries of knot complements. These relations are worked out in Chap. 9.

In Chaps. 7 and 8, we study the polyhedral decompositions for special classes of A-adequate or B-adequate links in more detail and obtain stronger versions of the main results.

In Chap. 10, we state several open questions and problems that have emerged from this work and discuss potential applications of the methods that we have developed.

Philadelphia, USA   D. Futer
East Lansing, USA   E. Kalfagianni
Provo, USA   J. Purcell

Acknowledgements

This manuscript is the fruit of a project conducted over several years, in several different locations, with the assistance of many people and organizations. We thank our home institutions, Temple University, Michigan State University, and Brigham Young University, for providing support and venues for various collaborative meetings on each of the three campuses and the National Science Foundation (NSF) for supporting this project through grant funding. During this project, Futer was supported by NSF grant DMS-1007221, Kalfagianni was supported by grants DMS-0805942 and DMS-1105843, and Purcell was supported by NSF grant DMS-1007437 and a Sloan Research Fellowship.

Some of this work was carried out and discussed while the authors were in attendance at workshops and conferences. These include the *Moab topology conference*, in Moab, Utah in May 2009; *Interactions between hyperbolic geometry, quantum topology, and number theory*, at Columbia University in June 2009; the *Joint AMS–MAA Meetings* in San Francisco in January 2010; *Topology and geometry in dimension three*, in honor of William Jaco, at Oklahoma State University in June 2010; *Knots in Poland III* at the Banach Center in Warsaw, Poland, in July 2010; and *Faces of geometry: 3-manifolds, groups, & singularities*, in honor of Walter Neumann, at Columbia University in June 2011. We thank the organizers of these conferences for their work and hospitality.

Our tools and techniques owe a sizable intellectual debt to the prior work of Ian Agol and Marc Lackenby. We are grateful to both of them for a number of enlightening discussions.

We thank Jessica Banks for her careful reading and numerous helpful suggestions. We also thank the MSU graduate students Cheryl Balm, Adam Giambrone, Christine Lee, and Indra Schottland for the interesting questions that they raised during a reading seminar on parts of this monograph.

We thank the referees for their helpful comments. We also thank Chris Atkinson, John Baldwin, Abhijit Champanerkar, Oliver Dasbach, Charlie Frohman, Cameron Gordon, Eli Grigsby, Joanna Kania-Bartoszynska, Ilya Kofman, Tao Li,

William Menasco, Walter Neumann, Neal Stoltzfus, and Roland van der Veen for their comments on, and interest in, this project.

Finally, we thank Yael Futer, George Pappas, and Tim Purcell for their support—especially during several extended visits that the authors had to pay at each other's households during the course of this project.

Contents

Chapter 1
Introduction

In the last 3 decades, there has been significant progress in 3-dimensional topology, due in large part to the application of new techniques from other areas of mathematics and from physics. On the one hand, ideas from geometry have led to geometric decompositions of 3-manifolds and to invariants such as the A-polynomial and hyperbolic volume. On the other hand, ideas from quantum physics have led to the development of invariants such as the Jones polynomial and colored Jones polynomials. While ideas generated by these invariants have helped to resolve several problems in knot theory, their relationships to each other, and to classical knot topology, are still poorly understood. Topological quantum field theory predicts that these invariants are in fact tightly related, as does mounting computer evidence. However, at this writing, several outstanding conjectures and open problems have been verified for only a handful of examples.

In this monograph, we initiate a systematic study of relations between quantum knot invariants and geometries of knot complements. We develop the setting and machinery that allows us to establish direct and concrete relations between colored Jones knot polynomials and geometric knot invariants. In several instances, our results provide deeper and more intrinsic explanations for the connections between geometry and quantum topology that have been observed in special cases in the past. In addition, this work leads to some surprising new relations between the two areas, and offers a promising environment for further exploring such connections.

We begin with some history and background on the problems under consideration, then give an overview of the work contained in this manuscript, including some of the results mentioned above.

1.1 History and Motivation

W. Thurston's ground-breaking work in the late 1970s established the ubiquity and importance of hyperbolic geometry in three-dimensional topology. In fact, hyperbolic 3-manifolds had been studied since the beginning of the twentieth

D. Futer et al., *Guts of Surfaces and the Colored Jones Polynomial*, Lecture Notes in Mathematics 2069, DOI 10.1007/978-3-642-33302-6_1,
© Springer-Verlag Berlin Heidelberg 2013

century as a subfield of complex analysis. In the 1960s and 1970s, Andreev
[7, 8], Riley [86, 87], and Jørgensen [51] found several families of hyperbolic
3-manifolds with increasingly complex topology. In particular, Riley constructed
the first examples of hyperbolic structures on complements of knots in the 3-sphere.
In a different direction, Jaco and Shalen [47] and Johannson [48] found a canonical
way to decompose a 3-manifold along surfaces of small genus (this is now called
the *JSJ decomposition* or *torus decomposition*). In particular, they observed that
simple 3-manifolds, i.e. ones that do not contain homotopically essential spheres,
disks, tori or annuli, have fundamental groups that share similar properties with
the groups of hyperbolic 3-manifolds. Thurston's major insight was that the pieces
of the JSJ decomposition should admit locally homogeneous geometric structures,
and furthermore that the simple pieces should admit complete hyperbolic structures.
This insight was formalized in the celebrated *geometrization conjecture*. Thurston
proved the conjecture for 3-manifolds with non-empty boundary [93], among others.
In 2003, Perelman proved the general conjecture [69, 79, 80].

A special case of Thurston's theorem [93] is that link complements in the
3-sphere satisfy the geometrization conjecture. In particular, the complement of
any non-torus, non-satellite knot must admit a complete hyperbolic metric. By
Mostow–Prasad rigidity [71, 82], this hyperbolic structure is unique up to isometry.
As a result, geometric information about a hyperbolic knot complement, such as
its volume, gives topological knot invariants. For arbitrary knots, one can obtain
a similar invariant, called the *simplicial volume*, by considering the sum of the
volumes of the hyperbolic components in the JSJ decomposition. The simplicial
volume is a constant multiple of the Gromov norm of the knot complement [44].

Since the mid-1980s, low-dimensional topology has also been invigorated by
ideas from quantum physics, which have led to powerful and subtle invariants.
The first major invariant along these lines is the celebrated Jones polynomial, first
formulated by Jones in 1985 using operator algebras [49]. Soon after, Kauffman
described a direct construction of the polynomial using the combinatorics of link
projections [55], and several authors generalized it to links and trivalent graphs
[28, 50, 56, 84]. Witten showed that the Jones polynomial of links in the 3-sphere
has an interpretation in terms of a $2 + 1$ dimensional *topological quantum field
theory* (TQFT). At the same time, he introduced new invariants for links in arbitrary
3-manifolds, as well as invariants of 3-manifolds [96, 97]. The resulting theory,
although defined only at the physical level of rigor, predicted that the Jones-type
invariants and their generalizations are intimately connected to geometric structures
of 3-manifolds, and particularly to hyperbolic geometry [96, p. 77]. As explained
by Atiyah [10], the TQFT proposed by Witten is completely characterized by
certain "gluing axioms." In the late 1980s, Reshetikhin and Turaev gave the first
mathematically rigorous construction of a TQFT that fit this axiomatic description
[85]. Unlike that of [97], which is intrinsically 3-dimensional, the constructions of
[85], as well as those of [55, 84], relied on combinatorial descriptions of 3-manifolds
and the representation theory of quantum groups. This approach makes it harder to
establish connections with the geometry of 3-manifolds.

In the 1990s, Kashaev defined an infinite family of complex valued invariants of links in 3-manifolds, using the combinatorics of triangulations and the quantum dilogarithm function [52]. For links in the 3-sphere, these invariants can also be formulated in terms of tangles and R-matrices [53]. Kashaev's invariants are parametrized by the positive integers; there is an invariant for each $n \in \mathbb{N}$. He conjectured that the large-n asymptotics of these invariants determine the volume of hyperbolic knots [54]. Building on these works, H. Murakami and J. Murakami were able to recover Kashaev's invariants as special values of the *colored* Jones polynomials: an infinite family of polynomials, closely related to the Jones polynomial, also parametrized by $n \in \mathbb{N}$ [73]. As a result, Kashaev's original conjecture has been reformulated into the *volume conjecture*, which asserts that the volume of a hyperbolic knot is determined by the large-n asymptotics of the colored Jones polynomials. Furthermore, Murakami and Murakami generalized the conjecture to all knots in S^3 by replacing the hyperbolic volume with the simplicial volume [73]. The volume conjecture fits into a more general, conjectural framework relating hyperbolic geometry and quantum topology; for details, see the survey papers [25, 72] and references therein. Despite compelling experimental evidence, the aforementioned conjectures are currently known for only a few examples of hyperbolic knots.

At the same time, a growing body of evidence points to strong relations between the coefficients of the Jones and colored Jones polynomials and the volume of hyperbolic links. One such form of evidence consists of numerical computations, for example those by Champanerkar, Kofman, and Paterson [18]. A second form of evidence consists of theorems proved for several classes of links, for example alternating links by Dasbach and Lin [23]. The authors of this monograph have extended those results to closed 3-braids [34], highly twisted links [32], and certain sums of alternating tangles [33]. The approach in all of these results is somewhat indirect, in that they relate hyperbolic volume to the Jones polynomial by estimating both quantities in terms of the twist number of a link diagram. To mention two examples, for alternating links the result follows from Lackenby's volume estimate in terms of the twist number in any alternating projections [58] and the relation of the twist number to the colored Jones polynomial observed by Dasbach and Lin [23]. For highly twisted links, our argument works as follows. First, we proved an effective version of Gromov and Thurston's 2π-theorem and applied it to estimate the hyperbolic link volume in terms of the twist number of any highly twisted projection. Second, we relied on the combinatorial properties of *Turaev surfaces*, as studied in [21], to relate the twist numbers to the coefficients of Jones polynomials. However, for general links, twist numbers have a highly imperfect relationship to hyperbolic volume [35]. This limits the applicability of these methods to special families of knots and links.

In this monograph, we modify our approach to these problems, focusing on the topology of incompressible surfaces in knot complements and their relations to the colored Jones knot polynomials. Our motivation for the project has been twofold. On the one hand, certain spanning surfaces of knots have been shown to carry information on colored Jones polynomials [21]. On the other hand, essential surfaces also shed light on volumes of manifolds [6] and additional geometry

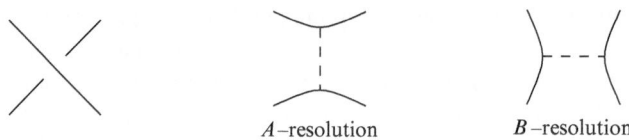

A–resolution   *B*–resolution

Fig. 1.1 *A*- and *B*-resolutions at a crossing of *D*

and topology (e.g. [2, 65, 68]). With these ideas in mind, we develop a machine that allows us to establish relationships between colored Jones polynomials and topological/geometric invariants.

For example, under mild diagrammatic hypotheses that arise naturally in the study of Jones-type polynomials, we show that the growth of the degree of the colored Jones polynomials is a *boundary slope* of an essential surface in the knot complement, as predicted by Garoufalidis [42]. Furthermore, certain coefficients of the polynomials measure how far this surface is from being a fiber in the knot complement. Our work leads to direct and detailed relations between hyperbolic geometry and Jones-type polynomials: for certain families of links, coefficients of the Jones and colored Jones polynomials determine the hyperbolic volume to within a factor of 4. Compared to previous arguments, which were all somewhat indirect, the way in which our machine produces volume inequalities gives a clearer and deeper conceptual explanation for why the hyperbolic volume should be related to particular coefficients of the Jones polynomial.

A survey of this monograph, in which the main theorems are illustrated by a running example, is given in [37].

1.2 State Graphs, and State Surfaces Far from Fibers

We begin with some terminology and conventions. Throughout this manuscript, $D = D(K)$ will denote a link diagram, in the equatorial 2-sphere of S^3. It is worth pointing out two conventions. First, we always assume (without explicit mention) that link diagrams are connected. Second, we abuse notation by referring to the projection 2-sphere using the common term *projection plane*. In particular, $D(K)$ cuts the projection "plane" into compact regions.

Let $D(K)$ be a (connected) diagram of a link K, as above, and let x be a crossing of D. Associated to D and x are two link diagrams, each with one fewer crossing than D, called the *A-resolution* and *B-resolution* of the crossing.

Definition 1.1. A *state* σ is a choice of *A*- or *B*-resolution at each crossing of D. Resolving every crossing, as in Fig. 1.1, gives rise to a crossing-free diagram $s_\sigma(D)$, which is a collection of disjoint circles in the projection plane. Thus one obtains a *state graph* $\mathbb{G}_\sigma$, whose vertices correspond to circles of s_σ and whose edges correspond to former crossings. For a given state σ, the *reduced state graph* $\mathbb{G}'_\sigma$ is the graph obtained from $\mathbb{G}_\sigma$ by removing all multiple edges between pairs of vertices.

The notion of states on link diagrams was first considered by Kauffman [55] during his construction of the bracket polynomial that provided a new construction and interpretation of the Jones polynomial.

Our primary focus is on the all-A and all-B states. The crossing-free diagram $s_A(D)$ is obtained by applying the A-resolution to each crossing of D. Its state graph is denoted $\mathbb{G}_A$ or $\mathbb{G}_A(D)$, and its reduced state graph $\mathbb{G}'_A$ or $\mathbb{G}'_A(D)$. Similarly, for the all-B state $s_B(D)$, the state graph is denoted $\mathbb{G}_B$, and the reduced state graph $\mathbb{G}'_B$.

To a state σ, we associate a *state surface* S_σ as follows. The state circles of σ bound disjoint disks in the 3-ball below the projection plane; these disks can be connected to one another by half-twisted bands at the crossings. The surface S_σ will have $\partial S_\sigma = K$. A special case of this construction is the Seifert surface constructed from the diagram $D(K)$, where the state σ is determined by an orientation on K.

When σ is the all-A or all-B state, the surfaces S_σ hold significance for both geometric topology and quantum topology. The graph $\mathbb{G}_A$ canonically embeds as a spine of the surface S_A. On the quantum side, the combinatorics of this embedding can be used to recover the colored Jones polynomials $J_K^n(t)$ [21, 23]. On the geometric side, as we will see below, the combinatorics of $\mathbb{G}_A$ dictates a geometric decomposition of the 3-manifold M_A obtained by cutting the link complement along the surface S_A. Because every statement has a B-state counterpart (by taking a mirror of the diagram), we will mainly discuss the all-A state for ease of exposition.

Definition 1.2. Let $M = S^3 \setminus K$ denote the 3-manifold with torus boundary component(s) obtained by removing a tubular neighborhood of K from S^3. Let S_A be the all-A state surface, as above, and let $M \backslash\backslash S_A$ denote the path-metric closure of $M \setminus S_A$. Note that $(S^3 \setminus K) \backslash\backslash S_A$ is homeomorphic to the 3-manifold $S^3 \backslash\backslash S_A$ obtained by removing a regular neighborhood of S_A from S^3. We will usually write $S^3 \backslash\backslash S_A$ for short, and denote this manifold with boundary by M_A.

We will refer to $P = \partial M_A \cap \partial M$ as the *parabolic locus* of M_A. This parabolic locus consists of annuli. The remaining, non-parabolic boundary $\partial M_A \setminus \partial M$ is the unit normal bundle of S_A.

Definition 1.3. Let M be an orientable 3-manifold and $S \subset M$ a properly embedded surface. We say that S is *essential* in M if the boundary of a regular neighborhood of S, denoted $\widetilde{S}$, is incompressible and boundary-incompressible. If S is orientable, then $\widetilde{S}$ consists of two copies of S, and the definition is equivalent to the standard notion of "incompressible and boundary-incompressible." If S is non-orientable, this is equivalent to π_1-injectivity of S, the stronger of two possible senses of incompressibility.

In the setting of Definition 1.2, the surface S_A is often non-orientable. In this case, $S^3 \backslash\backslash \widetilde{S_A}$ is the disjoint union of $M_A = S^3 \backslash\backslash S_A$ and a twisted I-bundle over S_A. Since we are interested in the topology of M_A, it is appropriate to look at the incompressibility of $\widetilde{S_A}$.

Guided by the combinatorial structure of the state graph $\mathbb{G}_A$, we construct a decomposition of M_A into topological balls. The connectivity properties of $\mathbb{G}_A$ govern the behavior of this decomposition; in particular, if $\mathbb{G}_A$ has no loop edges, we obtain a decomposition of M_A into checkerboard ideal polyhedra with 4-valent vertices (Theorem 3.12). This decomposition generalizes Menasco's decomposition of alternating link complements, which has been used frequently in the literature [64]. As a first application of our machinery, we use normal surface theory with respect to our polyhedral decomposition to give a new proof of the following theorem of Ozawa [76].

Theorem 3.19 (Ozawa). *Let $D(K)$ be a diagram of a link K. Then the all-A state surface S_A is essential in $S^3 \setminus K$ if and only if $\mathbb{G}_A$ contains no 1-edge loops. Similarly, the surface S_B is essential in $S^3 \setminus K$ if and only if $\mathbb{G}_B$ contains no 1-edge loops.*

Our polyhedral decomposition is designed to provide much more detailed information about the topology and geometry of $M_A = S^3 \backslash\backslash S_A$. In particular, we can characterize exactly when the surface S_A is a fiber of the link complement.

Theorem 5.11. *Let $D(K)$ be any link diagram, and let S_A be the spanning surface determined by the all-A state of this diagram. Then the following are equivalent:*

(1) The reduced graph $\mathbb{G}'_A$ is a tree.
(2) $S^3 \setminus K$ fibers over S^1, with fiber S_A.
(3) $M_A = S^3 \backslash\backslash S_A$ is an I-bundle over S_A.

It is remarkable to note that the state graph connectivity conditions that ensure incompressibility of the state surfaces first arose in the study of Jones-type knot polynomials. The following definition, formulated by Lickorish and Thistlethwaite [61,92], captures exactly the class of link diagrams whose polynomial invariants are especially well-behaved.

Definition 1.4. A link diagram $D(K)$ is called A-adequate (resp. B-adequate) if $\mathbb{G}_A$ (resp. $\mathbb{G}_B$) has no 1-edge loops. If both conditions hold for a diagram $D(K)$, then $D(K)$ and K are called *adequate*. If $D(K)$ is either A- or B-adequate, then $D(K)$ and K are called *semi-adequate*. As we will discuss in the next section, the hypothesis of semi-adequacy is rather mild.

Building on Theorem 5.11, we start with an A-adequate diagram D and strive to understand the geometric and topological complexity of $S^3 \backslash\backslash S_A$. In Chap. 2, we will see that the 3-manifold $M_A = S^3 \backslash\backslash S_A$ is in fact a handlebody, and thus atoroidal. The annulus version of the JSJ decomposition theory [47, 48] provides a way to cut M_A along annuli (disjoint from the parabolic locus) into three types of pieces: I-bundles over sub-surfaces of S_A, Seifert fibered spaces, and the *guts*, which is the portion that admits a hyperbolic metric with totally geodesic boundary. The Seifert fibered components are solid tori. Thus $\chi(\text{guts}(M_A)) = 0$ precisely when $\text{guts}(M_A) = \emptyset$ and M_A is a union of I-bundles and solid tori. In this case, M_A is called a *book of I-bundles* and S_A is called a *fibroid* [20]. The guts are the complex, interesting pieces of the geometric decomposition of M_A. Because hyperbolic surfaces, and guts, have negative Euler characteristic, it is convenient to work with the following definition.

Definition 1.5. Let Y be a compact cell complex, whose connected components are $Y_1, \ldots, Y_n$. Then the Euler characteristic of Y can be split into positive and negative parts:

$$\chi_+(Y) = \sum_{i=1}^{n} \max\{\chi(Y_i), 0\}, \qquad \chi_-(Y) = \sum_{i=1}^{n} \max\{-\chi(Y_i), 0\}.$$

It follows immediately that $\chi(Y) = \chi_+(Y) - \chi_-(Y)$. This notation is borrowed from the Thurston norm [94]. By convention, when $Y = \emptyset$, the above sums have no terms, hence $\chi_+(\emptyset) = \chi_-(\emptyset) = 0$.

The negative Euler characteristic $\chi_-(\text{guts}(M_A))$ serves as a useful measurement of how far S_A is from being a fiber or a fibroid in $S^3 \setminus K$. In fact, $\chi_-(\text{guts})$ is a key measurement of complexity in Agol's virtual fibering criterion [5], which is needed in the proof of the virtual fibering conjecture for hyperbolic 3-manifolds [4]. The Euler characteristic of guts also has a direct connection to hyperbolic geometry. Agol, Storm, and Thurston have shown that for any essential surface S in a hyperbolic 3-manifold M, a constant times $\chi_-(\text{guts}(M))$ gives a lower bound for $\text{vol}(M)$ [6]. This is applied below, in Sect. 1.5. On the other hand, the Euler characteristic $\chi(\mathbb{G}'_A)$ of the reduced graph $\mathbb{G}'_A$ first arose in the study of Jones-type polynomials [23,90], and in fact expresses one of their coefficients. This is explored in Sect. 1.4.

One of our main results is a diagrammatic formula for the guts of state surfaces for all A-adequate diagrams. In relating guts to reduced state graphs, it provides a bridge between hyperbolic geometry and quantum topology.

Theorem 5.14. *Let $D(K)$ be an A-adequate diagram, and let S_A be the essential spanning surface determined by this diagram. Then*

$$\chi_-(\text{guts}(S^3 \setminus\setminus S_A)) = \chi_-(\mathbb{G}'_A) - ||E_c||,$$

where $||E_c|| \geq 0$ is a diagrammatic quantity defined in Definition 5.9.

In many cases, the correction term $||E_c||$ vanishes. For example, this happens for alternating links [58], as well as for most Montesinos links. See Theorem 8.6, stated on p. 15 and Corollary 5.19 on p. 88. In each of these cases, Theorem 5.14 says that a geometric quantity, $\chi_-(\text{guts}(M_A))$, is equal to $\chi_-(\mathbb{G}'_A)$, which, as shown in [23], expresses a coefficient of the Jones polynomial.

1.3 Which Links are Semi-adequate?

We will be considering semi-adequate links throughout this manuscript. (After taking a mirror if necessary, such a link is A-adequate.) Before we continue with the description of our results, it is worth making some remarks about the class of semi-adequate links. It turns out that the class is very broad, and that the condition that a knot be semi-adequate seems to be rather mild. For example, with the exception of two 11-crossing knots that we will discuss below, and a handful of 12-crossings knots, all knots with at most 12 crossings are semi-adequate. Furthermore, every minimal crossing diagram for each of these semi-adequate knots is semi-adequate [90, 92]. Thus, apart from a few exceptions, our results in this monograph apply directly to the diagrams in the knot tables up to 12 crossings. The situation is similar with the larger tabulated knots: Stoimenow has computed that among the 253,293 prime knots with 15 crossings tabulated in [46], at least 249,649 are semi-adequate [91].

Several well-studied families of links are semi-adequate. These include alternating links, positive or negative closed braids, all closed 3-braids, all Montesinos links, and planar cables of all of the above. We refer the reader to [61, 91, 92] for more discussion and examples.

Nevertheless, there exist knots and links that are not semi-adequate. Before discussing examples, we recall that the Jones polynomial can be used to detect semi-adequacy. Indeed, the last coefficient of an A-adequate link must be ± 1. Similarly, the first coefficient of an B-adequate link must be ± 1 [92]. With the notation of Knotinfo [17], the knot $K = 11n_{95}$ has Jones polynomial equal to $J_K(t) = 2t^2 - 3t^3 + 5t^4 - 6t^5 + 6t^6 - 5t^7 + 4t^8 - 2t^9$. Hence, K is not semi-adequate; this is the first such knot in the knot tables. An infinite family of non semi-adequate knots, detected by the extreme coefficients of their Jones polynomial, can be obtained by [63, Theorem 5]. However, as we discuss below, the extreme coefficients of the Jones polynomial are not a complete obstruction to semi-adequacy.

Thistlethwaite [92] showed that certain coefficients of the 2-variable Kauffman polynomial [56] provide the obstruction to semi-adequacy. Building on Thistlethwaite's results, Stoimenow obtained a set of semi-adequacy criteria and applied them to several knots whose adequacy could not be determined by the Jones polynomial. For example, he showed that the knot $K' = 11n_{118}$ is not semi-adequate. Note that in this case, the last coefficient of the Jones polynomial, $J_{K'}(t) = 2t^2 - 2t^3 + 3t^4 - 4t^5 + 4t^6 - 3t^7 + 2t^8 - t^9$, is -1.

Ozawa has considered link diagrams and Kauffman states σ that are adequate (meaning $\mathbb{G}_\sigma$ has no 1-edge loops) and homogeneous (meaning $\mathbb{G}_\sigma$ contains a set of cut vertices that decompose it into a collection of all-A and all-B state graphs) [76]. See Definition 2.22 for more details. Semi-adequate diagrams clearly have this property, but the class of [76] is broader. As an example, consider the 12-crossing knot $K'' = 12n_{0706}$. This is not semi-adequate since both the extreme coefficients of the Jones polynomial are equal to 2. Indeed $J_{K''}(t) = 2t^{-4} - 4t^{-3} + 6t^{-2} - 8t^{-1} + 9 - 8t + 6t^2 - 4t^3 + 2t^4$. However K'' can be written as a 5-string braid that is homogeneous in the sense of Cromwell [19]. Thus the Seifert state of this closed braid diagram is homogeneous and adequate.

Ozawa proved that the state surface S_σ corresponding to a σ-adequate, σ-homogeneous diagram is always essential in $S^3 \setminus K$. In [29], Futer gave a direct proof of a slightly weaker version of Theorem 5.11, and also generalized it to σ-adequate, σ-homogeneous link diagrams. It turns out that many properties of the polyhedral decompositions that we develop below, as well as a number of results proved using the polyhedral decomposition, also extend to all adequate, homogeneous states. See Sects. 2.4, 3.4, 4.5, 5.6 where, in particular, we obtain analogues of Theorems 3.19, 5.11 and 5.14 in this generalized setting. Our study of the geometry of such links is continued in [31].

1.4   Essential Surfaces and Colored Jones Polynomials

The Jones and colored Jones polynomials have many known connections to the state graphs of diagrams. To specify notation, let

$$J_K^n(t) = \alpha_n t^{m_n} + \beta_n t^{m_n-1} + \ldots + \beta_n' t^{r_n+1} + \alpha_n' t^{r_n},$$

denote the n-th *colored Jones polynomial* of a link K. Recall that $J_K^2(t)$ is the usual Jones polynomial. Consider the sequences

$$js_K := \left\{ \frac{4m_n}{n^2} : n > 0 \right\} \quad \text{and} \quad js_K^* := \left\{ \frac{4r_n}{n^2} : n > 0 \right\}.$$

Garoufalidis' slope conjecture predicts that for each knot K, every cluster point (i.e., every limit of a subsequence) of js_K or js_K^* is a *boundary slope* of K [42], i.e. a fraction p/q such that the homology class $p\mu + q\lambda$ occurs as the boundary of an essential surface in $S^3 \setminus K$.

For a given diagram $D(K)$, there is a lower bound for r_n in terms of data about the state graph $\mathbb{G}_A(D)$, and this bound is sharp when $D(K)$ is A-adequate. Similarly, there is an upper bound on m_n in terms of $\mathbb{G}_B$ that is realized when $D(K)$ is B-adequate [60]. In [36], building on these properties and using Theorem 3.19, we relate the extreme degree of $J_K^n(t)$ to the boundary slope of S_A, as predicted by the slope conjecture.

Theorem 1.6 ([36]). *Let $D(K)$ be an A-adequate diagram of a knot K and let $b(S_A) \in \mathbb{Z}$ denote the boundary slope of the essential surface S_A. Then*

$$\lim_{n\to\infty} \frac{4r_n}{n^2} = b(S_A),$$

where r_n is the lowest degree of $J_K^n(t)$.

Similarly, if $D(K)$ is a B-adequate diagram of a knot K, let $b(S_B) \in \mathbb{Z}$ denote the boundary slope of the essential surface S_B. Then

$$\lim_{n\to\infty} \frac{4m_n}{n^2} = b(S_B),$$

where m_n is the highest degree of $J_K^n(t)$.

Work of Garoufalidis and Le [41, 43] implies that each coefficient of $J_K^n(t)$ satisfies linear recursive relations in n. For adequate links, these relations manifest themselves in a very strong form: Dasbach and Lin showed that if K is A-adequate, then the absolute values $|\beta_n'|$ and $|\alpha_n'|$ are independent of $n > 1$ [23]. In fact, $|\alpha_n'| = 1$ and $|\beta_n'| = 1 - \chi(\mathbb{G}_A')$, where $\mathbb{G}_A'$ is the reduced graph. Similarly, if D is B-adequate, then $|\alpha_n| = 1$ and $|\beta_n| = 1 - \chi(\mathbb{G}_B')$. Thus we can define the *stable values*

$$\beta_K' := |\beta_n'| = 1 - \chi(\mathbb{G}_A'), \qquad \text{and} \qquad \beta_K := |\beta_n| = 1 - \chi(\mathbb{G}_B').$$

The main results of this monograph explore the idea that the stable coefficient β_K' does an excellent job of measuring the geometric and topological complexity of the manifold $M_A = S^3 \backslash\backslash S_A$. (Similarly, β_K measures the complexity of $M_B = S^3 \backslash\backslash M_B$.) For instance, it follows from Theorem 5.11 that β_K' is exactly the obstruction to S_A being a fiber.

Corollary 9.16. *For an A-adequate link K, the following are equivalent:*

(1) $\beta_K' = 0$.
(2) For every A-adequate diagram of $D(K)$, $S^3 \setminus K$ fibers over S^1 with fiber the corresponding state surface $S_A = S_A(D)$.
(3) For some A-adequate diagram $D(K)$, $M_A = S^3 \backslash\backslash S_A$ is an I-bundle over $S_A(D)$.

Similarly, $|\beta_K'| = 1$ precisely when S_A is a fibroid of a particular type.

Theorem 9.18. *For an A-adequate link K, the following are equivalent:*

(1) $\beta_K' = 1$.
(2) For every A-adequate diagram of K, the corresponding 3-manifold M_A is a book of I-bundles, with $\chi(M_A) = \chi(\mathbb{G}_A) - \chi(\mathbb{G}_A')$, and is not a trivial I-bundle over the state surface S_A.
(3) For some A-adequate diagram of K, the corresponding 3-manifold M_A is a book of I-bundles, with $\chi(M_A) = \chi(\mathbb{G}_A) - \chi(\mathbb{G}_A')$.

In general, the geometric decomposition of M_A contains some non-trivial hyperbolic pieces, namely guts. In this case, $|\beta_K'|$ measures the complexity of the guts together with certain complicated parts of the maximal I-bundle of M_A. To state our result we need the following definition.

Definition 1.7. A link diagram D is called *prime* if any simple closed curve that meets the diagram transversely in two points bounds a region of the projection plane without any crossings.

Two crossings in D are defined to be *twist equivalent* if there is a simple closed curve in the projection plane that meets D at exactly those two crossings. The diagram is called *twist reduced* if every equivalence class of crossings is a *twist region* (a chain of crossings between two strands of K). The number of equivalence classes is denoted $t(D)$, the *twist number* of D.

Theorem 9.20. *Suppose K is an A-adequate link whose stable colored Jones coefficient is $\beta'_K \neq 0$. Then, for every A-adequate diagram $D(K)$,*

$$\chi_-(\text{guts}(M_A)) + ||E_c|| = |\beta'_K| - 1,$$

where as above $||E_c|| \geq 0$ is the diagrammatic quantity of Definition 5.9. Furthermore, if D is prime and every 2-edge loop in $\mathbb{G}_A$ has edges belonging to the same twist region, then $||E_c|| = 0$ and

$$\chi_-(\text{guts}(M_A)) = |\beta'_K| - 1.$$

To briefly discuss the meaning of the correction term $||E_c||$, recall that the non-hyperbolic components of the JSJ decomposition of M_A are I-bundles and solid tori. In Chap. 4, we show that the I-bundle components with negative Euler characteristic are spanned by *essential product disks (EPDs)*: properly embedded essential disks in M_A whose boundary meets the parabolic locus twice. These disks come in two types: those corresponding to (strings of) complementary regions of $\mathbb{G}_A$ with just two sides, and certain "complicated" ones, which we call *complex*. (See Definition 5.2 on p. 74.) The minimal number of complex EPDs in a spanning set is denoted $||E_c||$; this is exactly the correction term of Theorems 5.14 and 9.20.

It is an open question whether *every* A-adequate link admits a diagram for which $||E_c|| = 0$: see Question 10.2 on p. 156. For instance, Lackenby showed that this is the case for prime alternating links [58]. By Theorem 9.20, $||E_c|| = 0$ when every 2-edge loop of $\mathbb{G}_A$ has edges belonging to the same twist region. This is also the case for most Montesinos links (the reader is referred to Chap. 8 for the terminology).

Corollary 9.21. *Suppose K is a Montesinos link with a reduced admissible diagram $D(K)$ that contains at least three tangles of positive slope. Then*

$$\chi_-(\text{guts}(M_A)) = |\beta'_K| - 1.$$

Similarly, if $D(K)$ contains at least three tangles of negative slope, then

$$\chi_-(\text{guts}(M_B)) = |\beta_K| - 1.$$

When $||E_c|| = 0$, Theorem 9.20 offers striking evidence that coefficients of the colored Jones polynomials measure something quite geometric: when $|\beta'_K|$ is large, the link complement $S^3 \setminus K$ contains essential spanning surfaces that are correspondingly far from being a fiber. Whereas the Alexander polynomial and its generalization in Heegaard Floer homology are known to have many connections to the geometric topology of spanning surfaces of a knot [75, 77, 78], the geometric meaning of Jones-type polynomials has traditionally been a mystery. Theorems 9.16, 9.18, and 9.20 establish some of the first detailed connections between surface topology and the Jones polynomial.

1.5   Volume Bounds from Topology and Combinatorics

Recall that by the work of Agol, Storm, and Thurston [6], any computation of, or lower bound on, χ_-(guts) of an essential surface $S \subset S^3 \setminus K$ leads to a proportional lower bound on vol($S^3 \setminus K$). For instance, Lackenby's diagrammatic lower bound on the volumes of alternating knots and links came as a result of computing the guts of checkerboard surfaces [58]. However, computing χ_-(guts) has typically been quite hard: apart from alternating knots and links, there are very few infinite families of manifolds for which there are known computations of the guts of essential surface [3,57].

The results of this manuscript greatly expand the list of manifolds for which such computations exist. In Chap. 9, we combine [6] with our results in Theorems 5.14 and 9.20, as well as some of their specializations, to give lower bounds on hyperbolic volume for all A-adequate knots and links. See Theorem 9.3 on p. 140 for the most general result along these lines.

We also focus on two well-studied families of links: namely, positive braids and Montesinos links. For these families, we are able to compute or estimate the quantity $\chi_-(\text{guts}(M_A))$ in terms of much simpler diagrammatic data. As a consequence, we obtain tight, two-sided estimates on the volumes of knots and links in terms of the twist number $t(D)$ (see Definition 1.7).

Theorem 9.7. *Let $D(K)$ be a diagram of a hyperbolic link K, obtained as the closure of a positive braid with at least three crossings in each twist region. Then*

$$\frac{2v_8}{3}\, t(D) \;\leq\; \text{vol}(S^3 \setminus K) \;<\; 10v_3(t(D) - 1),$$

where $v_3 = 1.0149\ldots$ is the volume of a regular ideal tetrahedron and $v_8 = 3.6638\ldots$ is the volume of a regular ideal octahedron.

Observe that the multiplicative constants in the upper and lower bounds differ by a rather small factor of about 4.155. For Montesinos links, we obtain similarly tight two-sided volume bounds.

Theorem 9.12. *Let $K \subset S^3$ be a Montesinos link with a reduced Montesinos diagram $D(K)$. Suppose that $D(K)$ contains at least three positive tangles and at least three negative tangles. Then K is a hyperbolic link, satisfying*

$$\frac{v_8}{4}\,(t(D) - \#K) \;\leq\; \text{vol}(S^3 \setminus K) \;<\; 2v_8\, t(D),$$

where $v_8 = 3.6638\ldots$ is the volume of a regular ideal octahedron and $\#K$ is the number of link components of K. The upper bound on volume is sharp.

We also relate the volumes of these links to quantum invariants. Recall that the volume conjecture of Kashaev and Murakami–Murakami [54, 73] states that all hyperbolic knots satisfy

$$2\pi \lim_{n\to\infty} \frac{\log\left|J_K^n(e^{2\pi i/n})\right|}{n} = \mathrm{vol}(S^3 \setminus K).$$

If this volume conjecture is true, it would imply for large n a relation between the volume of a knot K and coefficients of $J_K^n(t)$. For example, for $n \gg 0$ one would have $\mathrm{vol}(S^3 \setminus K) < C\,||J_K^n||$, where $||J_K^n||$ denotes the L^1-norm of the coefficients of $J_K^n(t)$, and C is an appropriate constant. In recent years, a series of articles by Dasbach and Lin, as well as the authors, has established such relations for several classes of knots [24, 32–34]. In fact, in all known cases, the upper bounds on volume are paired with similar lower bounds. However, in all of the past results, showing that coefficients of $J_K^n(t)$ bound volume below required two steps: first, showing that Jones coefficients give a lower bound on twist number $t(D)$, and then showing that twist number gives a lower bound on volume. Each of these two steps is known to fail outside special families of knots [34,35], and their combination produces an indirect argument in which the constants are far from sharp.

By contrast, our results in this manuscript bound volume below in terms of a topological quantity, $\chi_-(\text{guts})$, that is directly related to colored Jones coefficients. As a consequence, we obtain much sharper lower bounds on volume, along with an intrinsic and satisfactory conceptual explanation for why these lower bounds exist. See Sect. 9.4 in Chap. 9 for more discussion.

Our techniques also imply similar results for additional classes of knots. For instance, Theorems 9.7 and 9.12 have the following corollaries.

Corollary 9.22. *Suppose that a hyperbolic link K is the closure of a positive braid with at least three crossings in each twist region. Then*

$$v_8\left(\left|\beta_K'\right| - 1\right) \leq \mathrm{vol}(S^3 \setminus K) < 15v_3\left(\left|\beta_K'\right| - 1\right) - 10v_3,$$

where $v_3 = 1.0149\ldots$ is the volume of a regular ideal tetrahedron and $v_8 = 3.6638\ldots$ is the volume of a regular ideal octahedron.

Corollary 9.23. *Let $K \subset S^3$ be a Montesinos link with a reduced Montesinos diagram $D(K)$. Suppose that $D(K)$ contains at least three positive tangles and at least three negative tangles. Then K is a hyperbolic link, satisfying*

$$v_8\left(\max\{|\beta_K|, |\beta_K'|\} - 1\right) \leq \mathrm{vol}(S^3 \setminus K) < 4v_8\left(|\beta_K| + |\beta_K'| - 2\right) + 2v_8\,(\#K),$$

where $\#K$ is the number of link components of K.

1.6 Organization

We now give a brief guide to the organization of this monograph.

In Chap. 2, we begin with a connected link diagram $D(K)$, and explain how to construct the state graph $\mathbb{G}_A$ and the state surface S_A. Guided by the structure of $\mathbb{G}_A$, we will cut the 3-manifold $M_A = S^3 \backslash\backslash S_A$ along a collection of disks into several topological balls. We obtain a collection of *lower* balls that are in one-to-one correspondence with the alternating tangles in $D(K)$ and a single *upper* 3-ball. The boundary of each ball admits a checkerboard coloring into white and shaded regions that we call faces. In the last section of the chapter we discuss the generalization of the decomposition to σ-homogeneous and σ-adequate diagrams.

In Chap. 3, we show that if $D(K)$ is A-adequate, each of these balls is a checkerboard colored ideal polyhedron with 4-valent vertices. This amounts to showing that the shaded faces on each of the 3-balls are simply-connected (Theorem 3.12). Furthermore, we show that the ideal polyhedra do not contain normal bigons (Proposition 3.18), which quickly implies Theorem 3.19. In the last section of the chapter, we generalize these results to homogeneous and adequate states.

In Chap. 4, we prove a structural result about the geometric decomposition of M_A. As already mentioned, the JSJ decomposition yields three kinds of pieces: I-bundles, solid tori, and the guts, which admit a hyperbolic metric with totally geodesic boundary. Let B be an I-bundle in the characteristic submanifold of M_A. We say that a finite collection of disjoint essential product disks (EPDs) $\{D_1, \ldots, D_n\}$ *spans* B if $B \setminus (D_1 \cup \cdots \cup D_n)$ is a finite collection of prisms (which are I-bundles over a polygon) and solid tori (which are I-bundles over an annulus or Möbius band). We prove the following.

Theorem 4.4. *Let B be a component of the characteristic submanifold of M_A which is not a solid torus. Then B is spanned by a collection of essential product disks (EPDs) $D_1, \ldots, D_n$, with the property that each D_i is embedded in a single polyhedron in the polyhedral decomposition of M_A.*

Like all results from the early chapters, Theorem 4.4 generalizes to σ-adequate and σ-homogeneous diagrams. See Sect. 4.5 for details.

In Chap. 5, we calculate the number of EPDs required to span the I-bundle of M_A. We do this by explicitly constructing a suitable spanning set of disks (Lemmas 5.6 and 5.8). The EPDs in the spanning set that lie in the *lower* polyhedra of the decompositions are well understood; they are in one-to-one correspondence with 2-edge loops in the state graph $\mathbb{G}_A$. The EPDs in the spanning set that lie in the upper polyhedron are *complex*; they are not parabolically compressible to EPDs in the lower polyhedra. The construction of this spanning set leads to a proof of Theorem 5.14. The spanning set of Chap. 5 also makes it straightforward to detect when the manifold M_A is an I-bundle, leading to a proof of Theorem 5.11.

The main tool used in Chaps. 3–5 is normal surface theory. In fact, our results about normal surfaces in the polyhedral decomposition of M_A can likely be used to attack other topological problems about A-adequate links: see Sect. 10.2 in Chap. 10 for variations on this theme.

The results of Chap. 5 reduce the problem of computing the Euler characteristic of the guts of M_A to counting how many complex EPDs are required to span the I-bundle of the *upper* polyhedron. In Chap. 6, we restrict attention to prime diagrams and address the problem of how to recognize such EPDs from the structure of the all-A state graph $\mathbb{G}_A$. Our main result there is Theorem 6.4, which describes the basic building blocks for such EPDs. Roughly speaking, each of these building blocks maps onto to a 2-edge loop of $\mathbb{G}_A$.

In Chap. 7, we restrict attention to A-adequate diagrams $D(K)$ for which the polyhedral decomposition includes no non-prime arcs or switches (see Definition 2.18 on p. 27). In this case, one can simplify the statement of Theorem 5.14 and give an easier combinatorial estimate for the guts of M_A. To state our result, let b_A denote the number of bigons in twist regions of the diagram such that a loop tracing the boundary of this bigon belongs to the B-resolution of D. (The A-resolution of these twist regions is *short* in Fig. 5.4 on p. 86.) Then, define $m_A = \chi(\mathbb{G}_A) - \chi(\mathbb{G}'_A) - b_A$. We prove the following estimate.

Theorem 7.2. *Let $D(K)$ be a prime, A-adequate diagram, and let S_A be the essential spanning surface determined by this diagram. Suppose that the polyhedral decomposition of $M_A = S^3 \backslash\backslash S_A$ includes no non-prime arcs. Then*

$$\chi_-(\mathbb{G}'_A) - 8m_A \;\leq\; \chi_-(\text{guts}(M_A)) \;\leq\; \chi_-(\mathbb{G}'_A),$$

where the lower bound is an equality if and only if $m_A = 0$.

In Chap. 8, we study the polyhedral decompositions of Montesinos links. The main result is the following.

Theorem 8.6. *Suppose K is a Montesinos link with a reduced admissible diagram $D(K)$ that contains at least three tangles of positive slope. Then*

$$\chi_-(\text{guts}(M_A)) = \chi_-(\mathbb{G}'_A).$$

Similarly, if $D(K)$ contains at least three tangles of negative slope, then

$$\chi_-(\text{guts}(M_B)) = \chi_-(\mathbb{G}'_B).$$

The arguments in Chaps. 6–8 require a detailed and fairly technical analysis of the combinatorial structure of the polyhedral decomposition; we call this analysis *tentacle chasing*. In addition, Chaps. 7 and 8 depend heavily on Theorem 6.4 in Chap. 6.

In Chap. 9, we give the applications to volume estimates and relations with the colored Jones polynomials that were discussed earlier in this introduction. The results in this chapter do not use Chap. 7 at all, and do not directly reference Chap. 6 or the arguments of Chap. 8. Thus, having the statement of Theorem 8.6 at hand, a reader who is eager to see the aforementioned applications may proceed to Chap. 9 immediately after Chap. 5.

In Chap. 10, we state several open questions and problems that have emerged from this work, and discuss potential applications of the methods that we have developed.

Chapter 2
Decomposition into 3-Balls

In this chapter, we start with a connected link diagram and explain how to construct state graphs and state surfaces. We cut the link complement in S^3 along the state surface, and then describe how to decompose the result into a collection of topological balls whose boundaries have a checkerboard coloring. There are two steps to this decomposition; the first is explained in Sect. 2.2, and the second in Sect. 2.3. Finally, in Sect. 2.4, we briefly describe how to generalize the decomposition to a broader class of links considered by Ozawa in [76].

The combinatorics of the decomposition will be used heavily in later chapters to prove our results. Consequently, in this chapter we will define terminology that will allow us to refer to these combinatorial properties efficiently. Thus the terminology and results of this chapter are important for all the following chapters.

2.1 State Circles and State Surfaces

Let D be a connected link diagram, and x a crossing of D. Recall that associated to D and x are two link diagrams, each with one fewer crossing than D, called the *A-resolution* and *B-resolution* of the crossing. See Fig. 1.1 on p. 4.

A *state* of D is a choice of A- or B-resolution for each crossing. Applying a state to the diagram, we obtain a crossing free diagram consisting of a disjoint collection of simple closed curves on the projection plane. We call these curves *state circles*. The *all-A state* of the diagram D chooses the A-resolution at each crossing. We denote the union of corresponding state circles by $s_A(D)$, or simply s_A. Similarly, one can define an all-B state and state circles $s_B = s_B(D)$.

Start with the all-A state of a diagram. From this, we may form a connected graph in the plane.

Definition 2.1. Let s_A be the union of state circles in the all-A state of a diagram D. To this union of circles, we attach one edge for each crossing, which records the location of the crossing. (These edges are dashed in Fig. 1.1 on p. 4.) The resulting

D. Futer et al., *Guts of Surfaces and the Colored Jones Polynomial*, Lecture Notes in Mathematics 2069, DOI 10.1007/978-3-642-33302-6_2,
© Springer-Verlag Berlin Heidelberg 2013

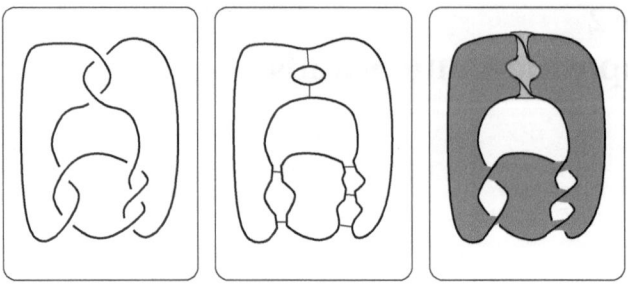

Fig. 2.1 *Left* to *right*: A diagram. The graph H_A. The state surface S_A

graph is trivalent, with edges coming from crossings of the original diagram and from state circles. To distinguish between these two, we will refer to edges coming from state circles just as state circles, and edges from crossings as *segments*. This graph will be important in the arguments below. We will call it *the graph of the A-resolution*, and denote it by H_A.

In the introduction, we introduced the *A-state graph* $\mathbb{G}_A$. This will factor into our calculations in later chapters. For now, note that $\mathbb{G}_A$ is obtained from H_A by collapsing state circles to single vertices.

We may similarly define the graph of the *B*-resolution, H_B, and the *B*-state graph $\mathbb{G}_B$. Indeed, every construction that follows will work with only minor modifications (involving handedness) if we replace *A*-resolutions with *B*-resolutions. For ease of exposition, we will mostly consider *A*-resolutions.

We now construct a surface related to a state σ. First, draw the circles of the σ-resolution, s_σ. These state circles bound disjoint disks in the 3-ball below the projection plane. Form the state surface S_σ by taking this disjoint collection of disks bounded by state circles, and attaching a twisted band for each crossing. The result is a surface whose boundary is the link. A well-known example of a state surface is the Seifert surface constructed from a diagram, where the state σ is chosen following an orientation on K.

See Fig. 2.1 for a state surface S_A corresponding to the all-A state.

Lemma 2.2. *The graph $\mathbb{G}_\sigma$ is a spine for the surface S_σ.*

Proof. By construction, $\mathbb{G}_\sigma$ has one vertex for every circle of s_σ (hence every disk in S_σ), and one edge for every half-twisted band in S_σ. This gives a natural embedding of $\mathbb{G}_\sigma$ into the surface, where every vertex is embedded into the corresponding disk, and every edge runs through the corresponding half-twisted band. This gives a spine for S_σ.   □

Lemma 2.3. *The surface S_σ is orientable if and only if $\mathbb{G}_\sigma$ is bipartite.*

Proof. It is well-known that a graph is bipartite if and only if all loops have even length.

For the "if" direction, assume that $\mathbb{G}_\sigma$ is bipartite. Then, we may construct a (transverse) orientation on S_σ, as follows. First, pick a normal direction to one disk (corresponding to one vertex of $\mathbb{G}_\sigma$). Then, extend over half-twisted bands to orient every adjacent disk, and continue inductively. This inductive construction of a transverse orientation on S_σ will never run into a contradiction, precisely because every loop in $\mathbb{G}_\sigma$ has even length. Thus S_σ is a two-sided surface in S^3, hence orientable.

For the "only if" direction, suppose $\mathbb{G}_\sigma$ is not bipartite, hence contains a loop of odd length. By embedding $\mathbb{G}_\sigma$ as a spine of S_σ, as in Lemma 2.2, we see that this loop of odd length is orientation-reversing on S_σ. □

In the special case where each state circle of σ traces a region of the diagram $D(K)$, the state graph $\mathbb{G}_\sigma$ is called a *checkerboard graph* or *Tait graph* and the surface Σ_σ is called a *checkerboard surface*. These checkerboard graphs record the adjacency pattern of regions of the diagram, and have been studied since the work of Tait. See e.g. [83, Page 264] for more discussion on the history of the subject.

Our primary focus will be on the all A-state and the A-state surface S_A. It will be helpful to isotope the state surface S_A into a topologically convenient position. Recall that by construction, the disks used to construct S_A lie in the 3-ball below the projection plane. Each of these disks can be thought of as consisting of a thin annulus with outside boundary attached to the state circle, and then a *soup can* attached to the inside boundary of the annulus. That is, a long cylinder runs deep under the projection plane, with a disk at the bottom. These soup cans will be nested, with outer state circles bounding deeper, wider soup cans. Finally, isotope the state circles and bands of the diagram onto the projection plane, except at crossings of the diagram in which the rectangular band runs through a small crossing ball coming out of the projection plane. When we have finished, aside from crossing balls, the link diagram sits on the plane of projection, and the surface S_A lies below.

Consider the manifold created by cutting S^3 cut along (a regular neighborhood of) S_A. We will refer to this manifold as $S^3 \backslash\backslash S_A$, or M_A for short. With S_A isotoped into the position above, it now is a straightforward matter to prove various topological conditions on M_A.

Lemma 2.4. *The manifold* $M_A = S^3 \backslash\backslash S_A$ *is homeomorphic to a handlebody.*

Proof. By definition, the manifold M_A is the complement of a regular neighborhood of S_A in S^3. A regular neighborhood of S_A consists of the union of a regular neighborhood of the link and a regular neighborhood of the twisted rectangles, as well as regular neighborhoods of each of the soup cans. Note first that the union of the regular neighborhood of the link and the rectangular bands deformation retracts to the projection graph of the diagram, which is a planar graph, and so its complement is a handlebody. Next, when we attach to this a regular neighborhood of a soup can, we are cutting the complement along a 2-handle. Since the result of cutting a handlebody along a finite number of non-separating 2-handles is still a handlebody, the lemma follows. □

2.2   Decomposition into Topological Balls

We will cut M_A along a collection of disks, to obtain a decomposition of the manifold into a collection of topological balls. In fact, we will eventually show this decomposition is actually a decomposition of M_A into ideal polyhedra, in the sense of the following definition.

Definition 2.5. An *ideal polyhedron* is a 3-ball with a graph on its boundary, such that complementary regions of the graph on the boundary are simply connected, and the vertices have been removed (i.e. lie at infinity).

Remark 2.6. Menasco's work [64] gives a decomposition of any link complement into ideal polyhedra. When the link is alternating, the resulting polyhedra have several nice properties. In particular, they are checkerboard colored, with 4-valent vertices. However, when the link is not alternating, these properties no longer hold. For alternating diagrams, our polyhedra will be exactly the same as Menasco's. More generally, we will see that our polyhedral decomposition of M_A also has a checkerboard coloring and 4-valent vertices.

There are two stages of the cutting. For the first stage, we will take one disk for each complementary region in the complement of the projection graph in S^2, with the boundary of the disk lying on S_A and the link. (Here we are using the assumption that our diagram is connected when we assert that the complementary regions of its projection graph are disks.) Note that each region of the projection graph corresponds to exactly one region of H_A, the graph of the A-resolution. Thus we may also refer to these disks as corresponding to regions of the complement of H_A in the projection plane.

To form the disk that we cut along, we isotope the disk of the given region by pushing it under the projection plane slightly, keeping its boundary on the state surface S_A, so that it meets the link a minimal number of times. Since S_A itself lies on or below the projection plane, except in the crossing balls, we know we can push the disk below the projection plane everywhere except possibly along the half-twisted rectangles at the crossings. At each crossing met by the particular region of H_A, the boundary of the region either runs past the twisted rectangular band without entering it, in which case it can be isotoped out of the crossing ball, or the boundary of the region runs along the attached band. In the latter case, the boundary comes into the crossing from below the projection plane. The crossing twists it such that if it continued to follow the band through the state surface, it would come out lying above the projection plane. To avoid this, isotope such that the boundary of the disk runs over the link inside the crossing ball, and so exits the crossing ball with the disk still under the projection plane.

After this isotopy, the result is one of the disks we cut along. We call such a disk a *white disk*, or *white face*, indicative of a checkerboard coloring we will give our polyhedral decomposition. Notice the above construction immediately gives the following lemma.

Fig. 2.2 A white disk lying just below the projection plane, with boundary (*dashed line*) on underside of *shaded* surface. Note this disk meets the link in exactly two points

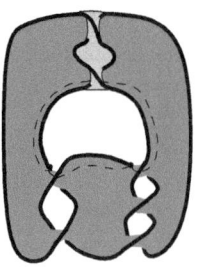

Lemma 2.7. *White disks meet the link only in crossing balls, and then only at under-crossings. Additionally, white disks lie slightly under the projection plane everywhere.* □

See Fig. 2.2 for an example.

Now, some of the white disks will not meet the link at all. These disks are those corresponding to regions of the projection graph whose boundaries never run through a crossing band. Therefore, the boundaries of such disks are isotopic to a state circle of s_A. Hence they are isotopic to soup cans attached to form the state surface. We call these particular soup cans *innermost disks*, since they will not have any additional soup cans nested inside them. Remove all white disks isotopic to innermost disks from consideration, since they are isotopic into the boundary of M_A.

We are left with a collection of disks $\mathscr{W}$, each lying in S^3 with boundary on the state surface S_A and on the link K. Cut along these disks.

Lemma 2.8. *Each component of $M_A \backslash\backslash \mathscr{W}$ is homeomorphic to a 3-ball.*

Proof. Notice there will be a single component above the projection plane. Since we have cut along each region of the projection graph, either by cutting along a soup can or along one of the disks in $\mathscr{W}$, this component must be homeomorphic to a ball.

Next, consider components which lie below the projection plane. These lie between soup can disks. Since any disk cuts the 3-ball below the projection plane into 3-balls, these components must also each be homeomorphic to 3-balls. □

Definition 2.9. The single 3-ball of the decomposition which lies above the plane of projection we call the *upper 3-ball*. All 3-balls below the plane of projection will be called *lower 3-balls*.

We now build up a combinatorial description of the upper and lower 3-balls. The boundary of any 3-ball will be marked by 2-dimensional regions separated by edges and ideal vertices. The regions (faces) come from white disks and portions of the surface S_A, which we shade.

Notation 2.10. In the sequel, we will use a variety of colors to label and distinguish the different shaded regions on the boundary of the 3-balls. All of these colored regions come from the surface S_A, and all of them are considered shaded. See, for example, Fig. 2.3 on p. 23.

Continuing the combinatorial description of the upper and lower 3-balls, edges on a 3-ball are components of intersection of white disks in $\mathcal{W}$ with (the boundary of a regular neighborhood of) S_A. Each edge runs between strands of the link. As usual, each ideal vertex lies on the torus boundary of the tubular neighborhood of a link component (see e.g. [64]).

Note each edge bounds a white disk in $\mathcal{W}$ on one side, and a portion of the shaded surface S_A on the other side. Thus, by construction, we have a checkerboard coloring of the 2-dimensional regions of our decomposition. Since the white regions are known to be disks, showing that our 3-balls are actually polyhedra amounts to showing that the shaded regions are also simply connected.

In the process of showing these regions are simply connected, we will build up a combinatorial description of how the white and shaded faces are super-imposed on the projection plane, and how these faces interact with the planar graph H_A. This combinatorial description will be useful in proving the main results.

Notation 2.11. From here on, we will refer to both white and shaded regions of our decomposition as *faces*. We do not assume that the shaded faces are simply connected until we prove they are, in Theorem 3.12.

Consider first the lower 3-balls.

Lemma 2.12. *Let R be a non-trivial component of the complement of s_A in the projection plane. Then there is exactly one lower 3-ball corresponding to R. The white faces of this 3-ball correspond to the regions in the complement of H_A that are contained in R.*

Here, by a non-trivial component of the complement of s_A, we mean a component which is not itself an innermost disk.

Proof. The soup cans attached to the circles s_A when forming S_A cut the 3-ball under the projection plane into the lower 3-balls of the decomposition. We will have exactly one such component for each non-trivial region of s_A. Faces are as claimed, by construction. $\qquad\square$

Lemma 2.13. *Each lower 3-ball is an ideal polyhedron, identical to the checkerboard polyhedron obtained by restricting to the alternating diagram given by the subgraph of H_A contained in a non-trivial region of s_A.*

Proof. Ideal edges of a lower 3-ball stretch from the link, across the state surface S_A, to the link again, and bound a disk of $\mathcal{W}$ on one side. The disks in $\mathcal{W}$, along which we cut, block portions of the link from view from inside the lower 3-ball. In particular, because each disk of $\mathcal{W}$ lies below the projection plane except at crossings, and the link lies on the projection plane except at crossings, the only parts of the link visible from inside a lower 3-ball correspond to crossings of the diagram. That is, only small segments of under-crossings of the link are visible from inside a lower 3-ball. Since edges meet at the same ideal vertex if and only if they meet the same strand of the link visible from below, edges of the lower 3-ball will meet other edges at under-crossings of the link. Notice that the only relevant under-crossings

Fig. 2.3 *Left* to *right*: An
example graph H_A. A
subgraph corresponding to a
region of the complement of
s_A. *White* and *shaded* faces of
the corresponding lower
polyhedron

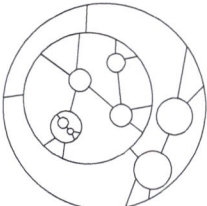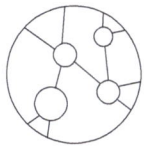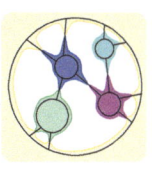

will be those which correspond to segments of H_A which lie inside the region R in the complement of s_A, as in Lemma 2.12. All other crossings are contained outside our 3-ball.

For each such under-crossing, note two disks of $\mathcal{W}$ meet at the under-crossing. These disks correspond to the two regions of H_A adjacent to the segment of H_A at that crossing, and they both meet in two edges. Thus each vertex is 4-valent. Finally, the graph formed by edges and ideal vertices must be connected, since the region of the complement of s_A is connected. Hence we have a 4-valent, connected graph on the plane, corresponding to a non-trivial subgraph of H_A contained in a single component of the complement of s_A.

Any connected, 4-valent graph on the plane corresponds to an alternating link, and gives the checkerboard decomposition of such a link. Thus it is an honest polyhedron. The vertices of the 4-valent graph correspond to crossings of the alternating diagram. Notice the vertices of the 4-valent graph also came from crossings of our link diagram. Thus we have a correspondence between a lower polyhedron and an alternating link with exactly the same crossings as in our subgraph of H_A.   □

By Lemma 2.13, we think of the lower polyhedra as corresponding to the largest alternating pieces of our knot or link diagram.

Schematically, to sketch a lower polyhedron, start by drawing a portion of H_A which lies inside a non-trivial region of the complement of s_A. Mark an ideal vertex at the center of each segment of H_A. Connect these dots by edges bounding white disks, as in Fig. 2.3.

Now we consider the upper 3-ball, or the single 3-ball lying above the plane of projection. Again, ideal edges on this 3-ball will meet at ideal vertices corresponding to strands of the link visible from inside the 3-ball. However, the identification no longer occurs only at single crossings. Still, we obtain the following.

Lemma 2.14. *The upper 3-ball admits a checkerboard coloring, and all ideal vertices are 4-valent.*

Proof. By shading the surface S_A gray and disks of $\mathcal{W}$ white, we obtain the checkerboard coloring.

Ideal vertices are strands of the link visible from inside the 3-ball. Each strand of the link is cut off as it enters an under-crossing. Thus visible portions of the link from above will lie between two under-crossings. At each under-crossing, two edges bounding a single disk meet the link on each side, as illustrated in Fig. 2.4.

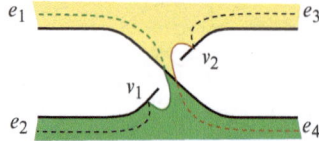

Fig. 2.4 Shown are portions of four ideal edges, terminating at under-crossings on a single crossing. Ideal edges e_1 and e_2 bound the same *white* disk and terminate at the ideal vertex v_1. Ideal edges e_3 and e_4 bound the same *white* disk and terminate at the ideal vertex v_2

Fig. 2.5 *Left*: A tentacle continues a *shaded* face in the upper 3-ball. *Right*: visualization of the tentacle on the graph H_A

Thus these two edges will share an ideal vertex with the two edges bounding a disk at the next under-crossing. Since no other edges meet the link strand between under-crossings, the vertex must be 4-valent.   □

We give a description of the upper 3-ball by drawing the faces, ideal edges, and ideal vertices of its boundary superimposed upon H_A, the graph of the A-resolution. We will see that the combinatorics of H_A determine the combinatorics of this upper 3-ball. We will continue referring to it as the "upper 3-ball" until we prove in Theorem 3.12 that it is indeed a polyhedron.

Notation 2.15. The combinatorial picture of the upper 3-ball superimposed on the graph of H_A will be described by putting together local moves that occur at each segment. In order to describe a particular move at a particular segment s, we will assume that the diagram has been rotated so that s is vertical. Now there are two state circles meeting s, one at the top of s and one at the bottom of s. Note the choice of top and bottom is not well-defined, but depends on how we rotated s to be vertical. However, the pair of edges of H_A, coming from state circles, which meet the segment at the top-right and at the bottom-left, is a well-defined pair.

Moreover, note that each crossing of the link diagram meets two shaded faces, one on the top and one on the bottom of the crossing, when the crossing is rotated to be as in Fig. 1.1. See also Fig. 2.4. Given a choice of shaded face, and a choice of crossing which that shaded face meets, this shaded face will determine a well-defined state circle meeting the segment s of H_A corresponding to that crossing. We may rotate H_A so that the segment s is vertical, and the state circle corresponding to our chosen shaded face is on the top, as in Fig. 2.5. Once we have performed this rotation, the state circle on top of s, as well as that on the bottom, and the four edges top-left, top-right, bottom-left, and bottom-right, are now completely well-defined.

The following description of the upper 3-ball, superimposed on the graph H_A, is illustrated in Figs. 2.5–2.7.

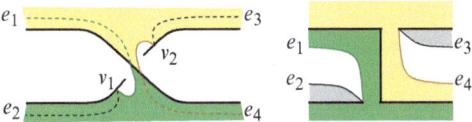

Fig. 2.6 The *shaded* faces around the crossing of Fig. 2.4 (Note: For grayscale versions of this monograph, *green* faces will appear as *dark gray*, *orange* faces as *lighter gray*)

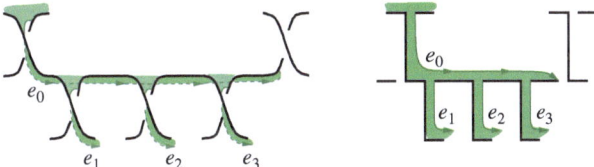

Fig. 2.7 *Left*: part of a *shaded* face in an upper 3-ball. *Right*: the corresponding picture, superimposed on H_A. The tentacle next to the ideal edge e_0 terminates at a segment on the same side of the state circle on H_A. It runs past segments on the opposite side of the state circle, spawning new tentacles associated to ideal edges e_1, e_2, e_3

First, we discuss ideal vertices of the upper 3-ball and how these appear on the graph H_A. Each ideal vertex corresponds to a strand of the diagram between two under-crossings, running over a (possibly empty) sequence of over-crossings. In terms of the graph H_A, for any given segment s, rotate H_A so that s is vertical, as in Notation 2.15. Now, draw two dots on the graph H_A on the edges at the top-right and bottom-left of s, near s. Erase a tiny bit of state circle between the segment s and each dot. After this erasure is performed at all segments of H_A, the connected components that remain (which are piecewise linear arcs between a pair of dots), correspond to ideal vertices. This erasing has been done on the top-right of Fig. 2.5, and for all edges shown in the right side of Fig. 2.7. It has also been done on the top-right and bottom-left of Fig. 2.6.

Every ideal edge of the upper 3-ball bounds a white disk on one side, and a shaded face on the other, and starts and ends at under-crossings. This ideal edge runs through a crossing, for which we may assume the orientation is as in Fig. 1.1, with the shaded face at the top of the crossing. Then, following Notation 2.15, this ideal edge will run from the top-right of a crossing, through the crossing, and continues parallel to the link strand on the bottom-right. Superimposed on H_A, the ideal edge will start at the top-right of a segment s, run adjacent to s, and then continue horizontally parallel to the state circle at the bottom of s. This is shown for a single edge in Fig. 2.5. Thus the white face adjacent to this ideal edge of the decomposition corresponds to the region of the complement of H_A to the right of this particular segment of H_A.

Note the shaded face adjacent to this ideal edge is the same shaded face on top of the segment of H_A, or from the top of the crossing. It runs between the link and the ideal edge, adjacent to the white face, until the ideal edge terminates at another under-crossing. We draw it on the graph H_A as shown in the right panel of Fig. 2.5.

Figure 2.7 shows multiple ideal edges. Figure 2.6 shows a single segment with the two ideal edges that are adjacent to it. Notice in that figure that when the green[1] shaded face is rotated to be on top, the green ideal edge runs from the top-right to the bottom-right of the segment. However, rotated as shown, the green runs from the bottom-left to the top-left. These are symmetric.

Definition 2.16. A *tentacle* is defined to be the strip of shaded face running from the top right of a segment of H_A, adjacent to the link, along the bottom right state circle of the edge of H_A, as illustrated in Fig. 2.5. Notice that a tentacle is bounded on one side by a portion of the graph H_A, and on the other side by exactly one ideal edge.

Note that a tentacle will continue past segments of H_A on the opposite side of the state circle without terminating, spawning new tentacles, but will terminate at a segment on the same side of the state surface. This is shown in Fig. 2.7.

Definition 2.17. A given tentacle is adjacent to a segment of H_A. Rotate the segment to be vertical, so that the tentacle lies to the right of this segment. The *head* of a tentacle is the portion attached to the top right of the segment of H_A. The *tail* is the part adjacent to the lower right of the state circle. We think of the tentacle as directed from head to tail.

Alternately, if we think of ideal edges of the decomposition as beginning at the top-right of a crossing, rotated as in Notation 2.15, and ending at the bottom-left, this orients each ideal edge. Since each tentacle is bounded by an ideal edge on one side, this in turn orients the tentacle. See Fig. 2.7.

Notice that for any segment, we will see the head of some tentacle at the top right of the segment, and the head of another tentacle at the bottom left of the segment. In Fig. 2.6, we see the heads of two tentacles. The orange one is on the top-right, the green one on the bottom-left. In addition, the tails of two other tentacles are shown in gray in that figure.

2.3 Primeness

Near the beginning of the last section, we stated that there were two stages to our polyhedral decomposition. The first stage is that explained above, given by cutting along white disks corresponding to complementary regions of H_A in the projection plane. This may not cut M_A into sufficiently simple pieces. In many cases, we may have to do some additional cutting to obtain polyhedra with the correct properties. This additional cutting is described in this section.

---

[1]Note: For grayscale versions of this monograph, green will refer to the darker gray shaded face, orange to the lighter gray.

Fig. 2.8 Arcs α_1, α_2, and α_3
are all non-prime arcs.
However, α_4 is not non-prime
in $H_A \cup (\alpha_1 \cup \alpha_2 \cup \alpha_3)$,
since there is a region in the
complement of
$H_A \cup (\alpha_1 \cup \cdots \cup \alpha_4)$ which
contains no state circles

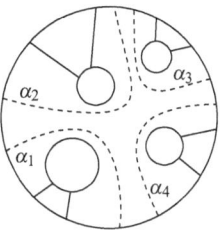

Recall from Definition 1.7 that a link diagram is *prime* if any simple closed curve
that meets the diagram transversely in two points bounds a region of the projection
plane with no crossings. By Lemma 2.13, the polyhedra in our decomposition that
lie below the projection plane correspond to alternating link diagrams. At this
stage, these diagrams may not all be prime. We need to modify the polyhedral
decomposition so that polyhedra below the projection plane do, in fact, correspond
to prime alternating diagrams.

Definition 2.18. The graph H_A is *non-prime* if there exists a state circle C of s_A
and an arc α with both endpoints on C such that the interior of α is disjoint from
H_A and α cuts off two non-trivial subgraphs when restricted to the subgraph of
H_A corresponding to the region of s_A containing α. The arc α is defined to be a
non-prime arc.

More generally, define a non-prime arc inductively as follows. Suppose
$\alpha_1, \ldots, \alpha_n$ are non-prime arcs with endpoints on the same state circle C. Suppose
there is a region R of the complement of $H_A \cup (\cup_{i=1}^n \alpha_i)$ and an arc β embedded in
that region with both endpoints on C such that β splits R into two new regions, both
containing state circles of s_A. Then $H_A \cup (\cup_{i=1}^n \alpha_i)$ is *non-prime*, and β is defined
to be a *non-prime arc*.

Figure 2.8 gives an example of a collection of non-prime arcs.

We call this non-prime because the corresponding alternating diagram of the
polyhedron below the projection plane will no longer be a prime alternating diagram
in the presence of such an arc.

The arc α meets two ideal edges of a polyhedron below the projection plane,
and two ideal edges of the 3-ball above the projection plane. Modify the polyhedral
decomposition as follows: we take our finger and push the arc α down against the
soup can corresponding to the state circle C. That is, we surger along the disk
bounded by α and an arc parallel to α running along the soup can. Topologically,
this divides the corresponding lower polyhedron into two, replacing two shaded
faces by one, and one white face by two. No new ideal vertices are added, but the
two ideal edges met by the non-prime arc are modified so that each runs from its
original head to a neighborhood of the non-prime arc, then parallel to the non-prime
arc, then along the tail of the other original ideal edge to where that ideal edge
terminates.

Combinatorially, this does the following. In the lower polyhedron, under the
projection plane, cut the polyhedron into two polyhedra by joining the ideal edges

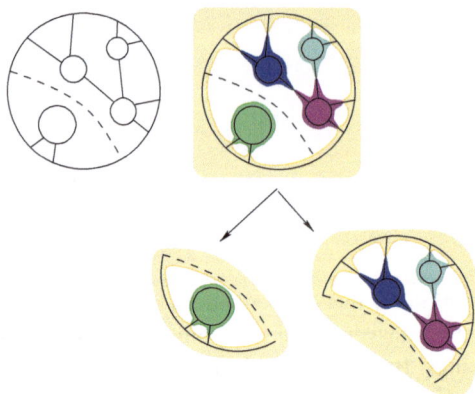

Fig. 2.9 Splitting a lower polyhedron into two along a non-prime arc. These give two polyhedral regions, defined in Definition 3.13 on p. 43

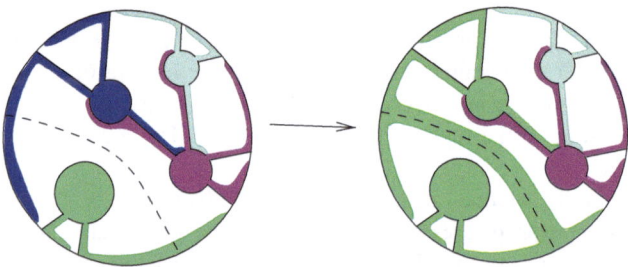

Fig. 2.10 Splitting the upper 3-ball along a non-prime arc

attached by the non-prime arc. See Fig. 2.9. The lower polyhedra now correspond to alternating links whose state circles contain α.

On the boundary of the upper 3-ball, connect tentacles at both endpoints of the non-prime arc α by attaching a small regular neighborhood of α, for example as in Fig. 2.10. We call this neighborhood of α connecting tentacles a *non-prime switch*. A priori, a non-prime switch might join two shaded faces into one, or else connect a shaded face to itself. (In fact, we will show in the next chapter that it does the former.) A non-prime switch also reroutes ideal edges adjacent to the connected tentacles to run adjacent to the non-prime arc. Notice that these two edges still have well defined orientations, although unlike the case of tentacles, this does not give a direction to the non-prime switch.

Definition 2.19. Let D be a diagram of a link, with H_A the corresponding graph of its A-resolution. Let $\alpha_1, \ldots, \alpha_n$ be a collection of non-prime arcs for H_A that is maximal, in the sense that each α_i is a non-prime arc in $H_A \cup (\cup_{j=1}^{i-1}\alpha_j)$ and there are no non-prime arcs in $H_A \cup (\cup_{j=1}^{n}\alpha_j)$. Cut M_A into upper and lower 3-balls along disks $\mathcal{W}$, as in Lemma 2.8. Then modify the decomposition by cutting lower polyhedra along each non-prime arc α_i, $i = 1, \ldots, n$, as described above. This decomposes M_A into 3-balls, which we continue to call upper and lower 3-balls. We refer to the decomposition as a *prime decomposition* of M_A.

Notice that the choice of a maximal collection of non-prime arcs may not be unique. In fact, by appealing to certain results about orbifolds, one can show that the pieces of the prime decomposition are unique. While we do not need this for our applications, the argument is outlined in Remark 3.17 on p. 45.

Definition 2.20. We say a polyhedron is *prime* if every pair of faces meet along at most one edge.

Equivalently, we will see that any prime polyhedron admits no normal bigons, as in Definition 3.16.

In our situation, we also have the following equivalent notion of *prime*. Recall that, by Lemma 2.13, each lower polyhedron corresponds to an alternating link diagram (which can be recovered from a sub-graph of H_A). The 4-valent graph of the polyhedron is identical to the 4-valent graph of the alternating diagram. Then the polyhedron will be prime if and only if the corresponding alternating diagram is prime, in the sense of Definition 1.7. This is one motivation for the notion of prime polyhedra.

The effect of the prime decomposition of M_A is summarized in the following lemma.

Lemma 2.21. *A prime decomposition of M_A, along a maximal collection of non-prime arcs $\alpha_1, \ldots, \alpha_n$, has the following properties:*

(1) *It decomposes M_A into one upper and at least one lower 3-ball.*
(2) *Each 3-ball is checkerboard colored with 4-valent vertices.*
(3) *Lower 3-balls are in one to one correspondence with non-trivial complementary regions of $s_A \cup (\cup_{i=1}^n \alpha_i)$.*
(4) *All lower 3-balls are ideal polyhedra identical to the checkerboard polyhedra of an alternating link, where the alternating link is obtained by taking the restriction of $H_A \cup (\cup_{i=1}^n \alpha_i)$ to the corresponding region of $s_A \cup (\cup_{i=1}^n \alpha_i)$, and replacing segments of H_A with crossings (using the A-resolution).*
(5) *The alternating diagram corresponding to each lower polyhedron is prime. Consequently, each lower polyhedron is itself prime.*
(6) *White faces of the 3-balls correspond to regions of the complement of $H_A \cup (\cup_{i=1}^n \alpha_i)$.*

Proof. Cutting along non-prime arcs may slice lower 3-balls into multiple pieces, but it will not subdivide the upper 3-ball. Hence we still have one upper and at least one lower 3-ball after a prime decomposition, giving item (1).

Note that ideal edges are modified by the prime decomposition, but each ideal edge still bounds a white face on one side and a shaded face on the other. Hence the 3-balls are still checkerboard colored. Moreover, the prime decomposition does not affect any ideal vertices, and so these remain 4-valent, as in Lemmas 2.13 and 2.14. This gives item (2).

For item (3), recall that before cutting along non-prime arcs, lower 3-balls corresponded to non-trivial regions of the complement of the state circles s_A. Now we cut these along non-prime arcs, splitting them into regions corresponding to components of the complement of $s_A \cup (\cup_{i=1}^n \alpha_i)$, as in Fig. 2.9.

Item (4) follows from Lemma 2.13 and from the fact that we cut along a maximal collection of non-prime arcs. Lower 3-balls were known to be ideal polyhedra corresponding to alternating links. When we cut along non-prime arcs, we modify the diagrams of these links by splitting into two along the non-prime arc. Because the collection of non-prime arcs is maximal, in the final result all such diagrams will be prime, proving (5).

Finally, before cutting along non-prime arcs, white faces corresponded to non-trivial regions in the complement of H_A. A non-prime arc will run through such a region, with its endpoints on the same state circle in the boundary of such a region. Hence after cutting along a non-prime arc, we have separated such a region into two. Item (6) follows.   □

At this stage, we have quite a bit of information about the lower 3-balls: we know that each lower ball is an ideal polyhedron, and that it is prime. The same statements are true for the upper 3-ball as well, although they are harder to prove. Proving these results for the upper 3-ball is one of the main goals of the next chapter.

2.4 Generalizations to Other States

So far, we have described how to decompose the surface complement $M_A = S^3 \backslash\backslash S_A$ into 3-balls. By reflecting a B-adequate diagram to make it A-adequate, one could apply the same decomposition to $S^3 \backslash\backslash S_B$. In this section, we briefly describe how to generalize the decomposition into 3-balls to the much broader class of σ-homogeneous states considered by Ozawa in [76].

Given a state σ of a link diagram $D(K)$, recall that s_σ is a collection of disjointly embedded circles on the projection plane. We obtain a trivalent graph H_σ by attaching edges, one for each crossing of the original diagram $D(K)$. The edges of H_σ that come from crossings of the diagram are referred to as *segments*, and the other edges are portions of state circles.

Definition 2.22. Given a state σ of a link diagram $D(K)$, the circles of s_σ divide the projection plane into components. Within each such component, we have a collection of segments coming from crossings of the diagram. Label each segment A or B, depending on whether the corresponding crossing is given an A or B-resolution in the state σ. If all edges within each component have the same A or B label, we say that σ is a *homogeneous* state, and the diagram D is σ-*homogeneous*. See Fig. 2.11 for an example.

Let G_σ be the graph obtained by collapsing the circles of H_σ into vertices. If G_σ contains no loop edges, we say that σ is an *adequate* state, and the diagram that gave rise to this state is σ-*adequate*.

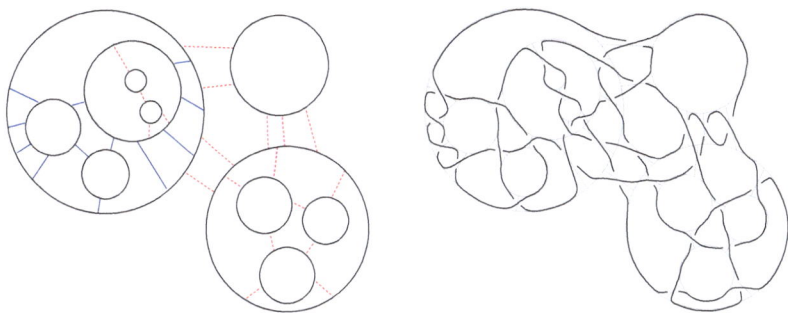

Fig. 2.11 An example of a σ-homogeneous diagram (on the *right*) and its graph H_σ (*left*). *Blue* segments represent the B-resolution, *red* segments the A-resolution. Note σ-homogeneity means each component in the complement of the state circles (*black circles on left*) has all segments of the same color

Let σ be a homogeneous state of a link diagram $D(K)$, and let S_σ denote the corresponding state surface. We let $M = S^3 \setminus K$ denote the link complement, and let $M_\sigma := M \backslash\backslash S_\sigma$ denote the path-metric closure of $M \setminus S_\sigma$. Note that $M_\sigma = (S^3 \setminus K) \backslash\backslash S_\sigma$ is homeomorphic to $S^3 \backslash\backslash S_\sigma$, obtained by removing a regular neighborhood of S_σ from S^3. As above, we will refer to $P = \partial M_\sigma \cap \partial M$ as the *parabolic locus* of M_σ; it consists of annuli.

We cut $M_\sigma = S^3 \backslash\backslash S_\sigma$ along disks, one for each region of the complement of H_σ, excepting innermost circles of s_σ. We refer to these disks as *white disks*, and we denote the collection of such disks by $\mathscr{W}$. This cuts M_σ into 3-balls: one *upper* 3-ball lying above the plane of projection, and multiple *lower* 3-balls, one for each component of s_σ. On the surface of each 3-ball is a graph, with edges coming from intersections of white disks and the surface S_σ, dividing the surface of the 3-ball into regions: *white faces*, coming from the white disks, and *shaded regions*, coming from portions of the state surface S_σ.

Because the diagram $D(K)$ is σ-homogeneous, each lower 3-ball comes from a sub-diagram of D that consists of only A- or only B-resolutions. Because this sub-diagram is contained in a single non-trivial component of the complement of the state circles s_σ, it is alternating. The proofs of Lemmas 2.12 and 2.13 go through in the σ-homogeneous setting, and we immediately obtain analogous results for the lower 3-balls in this case.

Lemma 2.23. *Let σ be a homogeneous state of a diagram $D(K)$. Let R be a non-trivial component of the complement of s_σ in the projection plane. Then:*

(1) There is exactly one lower 3-ball corresponding to R. Its white faces correspond to the regions in the complement of H_σ that are contained in R.

(2) Each lower 3-ball is an ideal polyhedron, identical to the checkerboard polyhedron obtained by restricting to the alternating diagram given by the subgraph of H_σ contained in a non-trivial region of s_σ. $\square$

As above, we call R a *polyhedral region*.

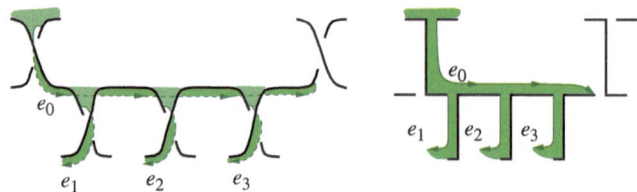

Fig. 2.12 Analogue of Fig. 2.7. When resolutions switch from all-A to all-B across a separating state circle of s_σ, we attach *left-down* tentacles rather than *right-down*

The ideal edges of the upper 3-ball are given by the intersection of white disks with the surface S_σ. Since each white disk is contained in a single polyhedral region R in the complement of the state circles s_σ, each crossing that the white disk borders has been assigned the same resolution, A or B, by σ. Thus the local description of these ideal edges is identical to that in the all-A or all-B case. In particular, obtain the analogue of Lemma 2.14: the upper 3-ball is checkerboard colored, and all ideal vertices are 4-valent. In the σ-homogeneous setting, the definitions of *tentacles* and their *head* and *tail* directions in the σ-homogeneous case, are completely analogous to Definitions 2.16 and 2.17.

Definition 2.24. For any segment of H_σ, rotate H_σ so that the segment is vertical. A *tentacle* is defined to be the strip of shaded face running from the top of the segment, adjacent to the link, along the bottom state circle adjacent to this segment. When the segment comes from a crossing with the A-resolution, the tentacle runs from the top to the right. When the segment comes from a crossing with the B-resolution, the tentacle runs from the top and to the left. A tentacle is bounded on one side by a portion of the graph H_σ, and on the other side by exactly one ideal edge.

The *head* of the tentacle is the portion attached to the top of the segment. The tail is adjacent to the state circle. We think of a tentacle as directed from head to tail.

In the σ-homogeneous setting, some tentacles run in the right-down direction (corresponding to the A-resolution) and some in the left-down direction (corresponding to the B-resolution). However, within any component of the complement of s_σ, all tentacles run in the same direction. Thus the only way to switch from right-down to left-down, or vice versa, is to cross over a circle of s_σ.

In the upper 3-ball, we attach tentacles to tentacles across state circles as shown in Fig. 2.7. However, if the state circle separates A-resolutions from B-resolutions, we attach left-down tentacles to right-down tentacles, or vice-versa. See Fig. 2.12.

Finally, to complete the decomposition, just as in the all-A or all-B case we need to ensure primeness of the polyhedra. To do so, we add a maximal collection of non-prime arcs, defined exactly as in Sect. 2.3, and then surger our polyhedra along disks bounded by these arcs. Because non-prime arcs connect a state circle to itself, and therefore separate only all-A or all-B resolutions (by σ-homogeneity), all the discussion in Sect. 2.3 goes through without modification in the σ-homogeneous case (except to replace all A's with all B's if necessary, which does not affect the argument).

Chapter 3
Ideal Polyhedra

Recall that $M_A = S^3 \backslash\backslash S_A$ is S^3 cut along the surface S_A. In the last chapter, starting with a link diagram $D(K)$, we obtained a prime decomposition of M_A into 3-balls. One of our goals in this chapter is to show that, if $D(K)$ is A-adequate (see Definition 1.1 on p. 4), each of these balls is a checkerboard colored ideal polyhedron with 4-valent vertices. This amounts to showing that the shaded faces on each of the 3-balls are simply-connected, and is carried out in Theorem 3.12.

Once we have established the fact that our decomposition is into ideal polyhedra, as well as a collection of other lemmas concerning the combinatorial properties of these polyhedra, two important results follow quickly. The first is Proposition 3.18, which states that all of the ideal polyhedra in our decomposition are prime. The second is a new proof of Theorem 3.19, originally due to Ozawa [76], that the surface S_A is essential in the link complement if and only if the diagram of our link is A-adequate.

All the results of this chapter generalize to σ-adequate, σ-homogeneous diagrams. We discuss this generalization in Sect. 3.4.

The results of this chapter will be assumed in the sequel. To prove many of these results, we will use the combinatorial structure of the polyhedral decomposition of the previous chapter, in a method of proof we call *tentacle chasing*. This method of proof, as well as many lemmas established here using this method, will be used again quite heavily in parts of Chaps. 4, 6–8. Therefore, the reader interested in those chapters should read the tentacle chasing arguments carefully, to be prepared to use such proof techniques later. In particular, tentacle chasing methods form a crucial component in the proofs of our main results, which reside in Chaps. 5 and 9 respectively.

However, a reader who is eager to get to the main theorems and their applications, and who seeks only a top-level outline of the proofs, may opt to survey the results of this chapter while taking the proofs on faith. The top-level proofs of the main results in Chap. 5 and the applications in Chap. 9 will not make any direct reference to tentacle chasing.

D. Futer et al., *Guts of Surfaces and the Colored Jones Polynomial*, Lecture Notes in Mathematics 2069, DOI 10.1007/978-3-642-33302-6_3,

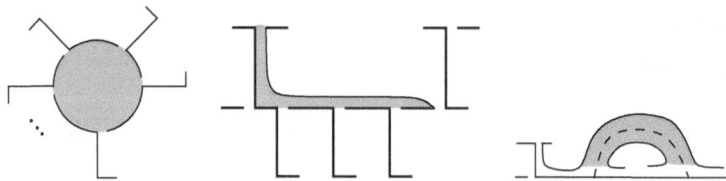

Fig. 3.1 Building blocks of a shaded face: an innermost disk, a tentacle, and a non-prime switch

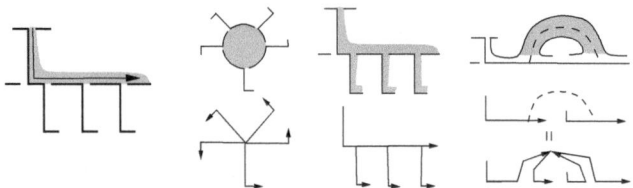

Fig. 3.2 *Far left*: A directed spine of a tentacle. *Left* to *right*: Shown is how directed tentacles connect to an innermost disk, to another tentacle, across a non-prime switch

3.1   Building Blocks of Shaded Faces

To prove the main results of this chapter, first we need to revisit our construction of shaded faces for the upper 3-ball. Shaded faces in the upper 3-ball are built of one of three pieces: innermost disks, tentacles, and non-prime switches. See Fig. 3.1. Recall that a tentacle is directed, starting at the portion adjacent to the segment of H_A (the head) and ending where the tentacle terminates adjacent to the state circle (the tail). This direction leads naturally to the definition of a *directed spine* for any shaded face on the upper 3-ball, as follows. For each tentacle, take a directed edge running through the core of the tentacle, with initial vertex on the state circle to which the segment of the tentacle is attached, and final vertex where the tentacle terminates, adjacent to the state circle. For each innermost disk, take a vertex. Notice that innermost disks are sources of directed edges of the spine, with one edge running out for each segment adjacent to the disk, but no directed edges running in. A non-prime arc is also represented as a vertex of the spine, with two incoming edges and two outgoing edges. This motivates the term *non-prime switch*. See Fig. 3.2.

In the language of directed spines, the statement that shaded faces are simply connected (Theorem 3.12) can be rephrased to say that the directed spine of each shaded face is, in fact, a directed tree.

Definition 3.1. When an oriented arc running through a tentacle in a shaded face is running in the same direction as that of the orientation above, or in the same direction as the edge of the directed spine, we say the path is running *downstream*. When the oriented path is running opposite the direction on the tentacle, we say the path is running *upstream*.

Figure 3.2, far left, shows an arc running through a single tentacle in the downstream direction. All the arrows in the remainder of that figure point in the downstream direction.

Definition 3.2. Suppose a directed arc γ, running through a shaded face of the upper 3-ball, has been homotoped to run monotonically through each innermost disk, tentacle, and non-prime switch it meets. Suppose further that γ meets any innermost disk, tentacle, and non-prime switch at most once. Then we say that γ is *simple with respect to the shaded face*.

Note that paths through the spine of a shaded face are simple if and only if they are embedded on the spine.

We say that γ is *trivial* if it does not cross any state circles.

3.2 Stairs and Arcs in Shaded Faces

The directions given to portions of shaded faces above lead to natural directions on subgraphs of H_A. One subgraph of H_A that we will see repeatedly is called a right-down staircase.

Definition 3.3. A *right-down staircase* is a connected subgraph of H_A determined by an alternating sequence of state circles and segments of H_A, oriented so that every turn from a state circle to a segment is to the right, and every turn from a segment to a state circle is to the left. (So the portions of state circles and edges form a staircase moving down and to the right.)

In fact, right-down staircases could be named left-up, except that the down and right follows the convention of Notation 2.15.

In this section, we present a series of highly useful lemmas that will allow us to find particular right-down staircases in the graph H_A associated with shaded faces. These lemmas lead to the proof of Theorem 3.12, and will be referred to frequently in Chaps. 4, 6–8.

Lemma 3.4 (Escher stairs). *In the graph H_A for an A-adequate diagram, the following are true:*

(1) *No right-down staircase forms a loop, and*
(2) *No right-down staircase has its top and bottom on the same state circle.*

Cases (1) and (2) of Lemma 3.4 are illustrated in Fig. 3.3.

Proof. Suppose there exists a right-down staircase forming a loop. Notice that the staircase forms a simple closed curve in the projection plane. Each state circle of the staircase intersects that loop. Because state circles are also simple closed curves, they must intersect the loop an even number of times. Because state circles cannot intersect segments, each state circle within the loop must be connected to another state circle within the loop. There must be an outermost such connection. These

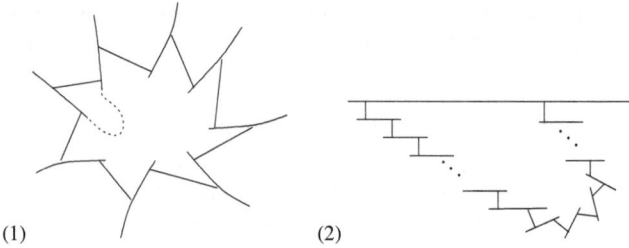

Fig. 3.3 *Left*: A right-down staircase forming a loop. *Right*: A single *right-down* staircase with its *top* and *bottom* connected to the same state circle

two state circles will form adjacent stairs, and connect within the loop. But then the segment between them gives a segment with both endpoints on the same state circle, contradicting A-adequacy of the diagram, Definition 1.4 (p. 6).

Similarly, suppose a right-down staircase has its top and bottom on the same state circle. Then the staircase and this state circle forms a loop, as above, and state circles that enter the loop must connect to each other. Again there must be some outermost connected pair. This pair will be two adjacent stairs. Again the segment between them will then give a segment with both endpoints on the same state circle, contradicting A-adequacy.   □

Lemma 3.4 is the first place where we have used A-adequacy. In fact, as the following example demonstrates, this hypothesis (or a suitable replacement, such as σ-adequacy) is crucial for both the lemma and for future results.

Example 3.5. Consider the unique connected, two-crossing diagram of a two-component unlink. This diagram is not A-adequate. Its graph H_A features both a loop staircase (with two steps), and a one-step staircase with its top and bottom on the same state circle, violating both conclusions of Lemma 3.4.

The loop staircase also gives rise to a non-trivial loop in the directed spine of the (unique) shaded face. Thus the upper 3-ball of this diagram is not a polyhedron. Therefore, all the proof techniques requiring a polyhedral decomposition will fail for this inadequate diagram.

Definition 3.6. Every non-prime arc α_i has its endpoints on some state circle C, and cuts a disk in the complement of C into two regions, called *non-prime half-disks*.

The following lemma will help us deal with combinatorial behavior when we encounter non-prime arcs.

Lemma 3.7 (Shortcut lemma). *Let α be a non-prime arc with endpoints on a state circle C. Suppose a directed arc γ lies entirely on a single shaded face, and is simple with respect to that shaded face, in the sense of Definition 3.2. Suppose γ runs across α into the interior of the non-prime half-disk bounded by α and C, and then runs upstream. Finally, suppose that γ exits the interior of that half-disk across the state* *circle C. Then γ must exit by following a tentacle downstream (that is, it cannot exit running upstream).*

Proof. Consider an innermost counterexample. That is, if there exists a counterexample, then there exists one for which γ does not cross any other non-prime arc and then run upstream when exiting the non-prime half-disk bounded by C and α. Consider the subarc of γ which runs from the point where it crosses α to the point where it crosses C. We will abuse notation slightly and call this arc γ.

After crossing α, the arc γ is running upstream in a tentacle adjacent to C. Note that since we are assuming this is a counterexample, it will not cross C immediately, for to do so it would follow a tentacle running downstream. Additionally, it cannot cross some other non-prime arc α_1 with endpoints on C, for because we are assuming this counterexample is innermost, it would then exit the region bounded by α_1 and C running downstream, contradicting our assumption that it crosses C running upstream. Finally, it may reach a non-prime arc α_1 and run around it without crossing, but then we are still running upstream on a tentacle adjacent to C, so we may ignore this case.

Hence the only possibility is that γ crosses α and then runs up the head of a tentacle with tail on C. The head of this tentacle is adjacent to a single step of a right-down stair. Consider what γ may do at the top of this stair.

(1) It may continue upstream, following another tentacle.
(2) It may change direction, following a tentacle downstream, or crossing a non-prime arc α_1 with endpoints on C_1 and then (eventually) running downstream across C_1.
(3) It may run over a non-prime switch without crossing the non-prime arc.

By assumption (counterexample is innermost), it cannot run over a non-prime arc α_1 with endpoints on C_1 and (eventually) cross C_1 running upstream. Notice that if γ enters an innermost disk, it must leave the disk running downstream, case (2), since an innermost disk is a source for edges of the directed spine. Also, in case (3), γ remains adjacent to the same state circle before and after, and so we ignore this case.

In case (1), we follow γ upstream to a new stair, and the same options are again available for γ, so we may repeat the argument.

We claim that γ is eventually in case (2). For, suppose not. Then since γ crosses C, and the graph H_A is finite, by following tentacles upstream we form a finite right-down staircase whose bottom is on C, and whose top is on C as well. This contradicts Lemma 3.4 (Escher stairs).

So eventually γ must change direction, following a tentacle downstream. After following the tentacle downstream, γ will be adjacent to another state circle. At this point, it may do one of two things:

(1) It may continue downstream through another tentacle, or by running through a non-prime arc first and then continuing downstream.
(2) It may run over a non-prime switch without crossing the non-prime arc.

Notice that these are the only options because first, no arc running downstream can enter an innermost disk (because such a disk is a source). Second, by assumption (innermost) γ cannot cross a non-prime arc and then cross the corresponding state circle running upstream. Third, tentacles only connect to tentacles in a downstream direction (Fig. 3.2 center). Again we ignore case (2), as γ will be adjacent to the same state circle before and after running over the non-prime switch.

But since these are the only possibilities, γ must continue running downstream, and cannot change direction again to run upstream. Thus γ must exit C by running over a tentacle in the downstream direction. $\square$

Definition 3.8. The proof of the previous lemma involved following arcs through oriented tentacles, keeping track of local possibilities. We call this proof technique *tentacle chasing*. We will use it repeatedly in the sequel.

Lemma 3.9 (Staircase extension). *Let γ be a directed arc lying entirely in a single shaded face, such that γ is simple with respect to the shaded face (Definition 3.2). Suppose also that γ begins by crossing a state circle running downstream. Suppose that every time γ crosses a non-prime arc α with endpoints on C and enters the non-prime half-disk bounded by α and C, that it exits that half-disk. Then γ defines a right-down staircase such that every segment of the staircase is adjacent to γ, with γ running downstream. Moreover, the endpoints of γ lie on tentacles that are adjacent to the first and last stairs of the staircase.*

Proof. The arc γ runs through a tentacle downstream. The tentacle is attached to a state circle at its head, is adjacent to a segment of H_A, and then adjacent to a second state circle at its tail. Form the first steps of the right-down staircase by including the state circle at the head, the segment, and the state circle at the tail.

Now we consider where γ may run from here. Note it cannot run into an innermost disk, since each of these is a source (and so is entered only running upstream). Thus it must do one of the following:

(1) It runs through another tentacle downstream.
(2) It runs through a non-prime switch, without changing direction.
(3) It runs through a non-prime switch, changing direction.

In case (1), we extend the right-down staircase by attaching the segment and state circle of the additional tentacle. If γ continues, we repeat the argument with γ adjacent to this new state circle.

We ignore case (2), because γ will remain adjacent to the same state circle in this case, still running in the downstream direction.

In case (3), γ is adjacent to a state circle C, then enters a non-prime half-disk bounded by a non-prime arc and C. By hypothesis, γ also exits that half-disk. Since it cannot exit along the non-prime switch, by hypothesis that γ runs monotonically through non-prime switches and meets each at most once, γ must exit by crossing C. Then Lemma 3.7 implies that γ exits by following a tentacle downstream. This tentacle will be adjacent to some segment attached to C and a new state circle attached to the other endpoint of this segment. Extend the right-down staircase by

Fig. 3.4 Extend a *right-down* staircase over a non-prime switch

attaching this segment and state circle to C. See Fig. 3.4. If γ continues, we may repeat the argument.

After a finite number of repetitions, γ must terminate, and we have our extended right-down staircase as claimed in the lemma.   □

The following is an immediate, highly useful consequence.

Lemma 3.10 (Downstream continues down, or Downstream lemma). *Let γ be as in Lemma 3.9. Then γ crosses the last state circle of the staircase by running downstream.*   □

We can now prove a result, which is called the Utility lemma because we will use it repeatedly in the upcoming arguments.

Lemma 3.11 (Utility lemma). *Let γ be a simple, directed arc in a shaded face, which starts and ends on the same state circle C. Then γ starts by running upstream from C, and then terminates at C while running downstream.*

Furthermore, γ cannot intersect C more than two times.

Proof. First, suppose that γ runs downstream from its first intersection with C. This will lead to a contradiction.

We begin by applying Lemma 3.9 (Staircase extension) to find a right-down staircase starting on C, such that γ runs downstream, adjacent to each segment of the staircase. This staircase will continue either until the terminal end of γ, or until γ crosses a non-prime arc α and enters (but does not exit) a half-disk R bounded by α and some state circle C'. But any such non-prime half-disk R will not contain the initial endpoint of γ (else γ would have crossed C' running downstream earlier, and we would have created a right-down staircase from C' to C', contradicting Lemma 3.4), hence R will not contain C unless $C' = C$. Because the final endpoint of γ is on C, either no such region R exists, or α has both endpoints on C. In either case, we will have constructed a right-down staircase that starts and ends on C, contradicting Lemma 3.4 (Escher stairs). So γ cannot run downstream from C.

Next, suppose that the terminal end of γ meets C running upstream. Then we simply reverse the orientation on γ, and repeat the above argument to obtain a contradiction. Therefore, γ first runs upstream from C, then terminates on C while running downstream.

Finally, suppose that γ meets C more than twice. Let $x_1, \ldots, x_n$ be its points of intersection with C. Applying the above argument to the sub-arc of γ from x_1 to x_2, we conclude that γ must arrive at x_2 while running downstream. But then the sub-arc of γ from x_2 to x_3 departs C running downstream, which is a contradiction.   □

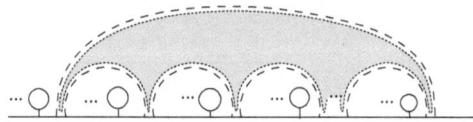

Fig. 3.5 If γ runs over a sequence of non-prime arcs, then γ bounds a region (shown shaded above) containing no state circles, giving a contradiction. Compare with Fig. 2.8

Given the above tools, we are now ready to show that our decomposition is into ideal polyhedra. The following is one of the main results of this chapter.

Theorem 3.12. *Let $D(K)$ be an A-adequate link diagram. Then, in the prime decomposition of M_A, shaded faces on the 3-balls are all simply connected. This gives a decomposition of M_A into checkerboard colored ideal polyhedra with 4-valent vertices.*

Proof. By Lemma 2.21, part (4), the lower 3-balls are ideal polyhedra, with simply connected faces. Hence, we need only consider the shaded faces on the upper 3-ball.

We have constructed a spine for each shaded face on the upper 3-ball. The shaded face will be simply connected if and only if the spine is a tree. Hence, we show the spine is a tree.

If the spine is not a tree, then there is a non-trivial embedded loop γ in the spine for the shaded face. Since γ is embedded in the spine, any sub-arc is simple in the sense of Definition 3.2.

Now, suppose γ crosses a state circle C. Since γ is a simple closed curve, as is the state circle, γ must actually cross C at least twice. Then we can express γ as the union of two directed arcs γ_1, γ_2, with endpoints at C, such that γ_1, γ_2 meet only at their endpoints. Suppose that both arcs are directed along a consistent orientation of γ. Then Lemma 3.11 (Utility lemma) says that γ_1 terminates at C running downstream. This means that γ_2 starts at C by running downstream, which contradicts the Utility lemma.

So γ never crosses a state circle. Since γ is non-trivial, contained in a single shaded face, it must run over a sequence of non-prime switches, all with endpoints on the same state circle C. When γ runs from one non-prime switch into another, it cannot meet any segments of H_A coming out of C, else the tentacle that γ runs through would terminate (γ would have to exit the shaded face). But then γ bounds a region in the projection plane which contains no state circles, since our diagram is assumed to be connected. This contradicts the definition of a collection of non-prime arcs, Definition 2.18 on page 27: the last such arc added to our collection divides a region of the complement of H_A and the other non-prime arcs into two pieces, one of which does not contain any state circles. See Fig. 3.5.

So shaded faces are simply connected. Since white faces are disks by definition, a prime decomposition of $M_A = S^3 \backslash\backslash S_A$ is a decomposition into ideal polyhedra. The fact that it is 4-valent and checkerboard colored follows from Lemma 2.21.   □

Recall that lower 3-balls are ideal polyhedra corresponding to non-trivial complementary regions of $s_A \cup (\cup \alpha_i)$, where the α_i form a maximal collection of non-prime arcs.

Definition 3.13. A *polyhedral region* is a complementary region of $s_A \cup (\bigcup \alpha_i)$ on the projection plane. With the convention that the "projection plane" is a 2-sphere, it follows that each polyhedral region is compact.

Lemma 3.14 (Parallel stairs). *Let σ_1 and σ_2 be simple, disjoint, directed arcs through the spines of shaded faces F_1 and F_2. (These shaded faces are allowed to coincide, so long as the σ_i are disjoint.) Suppose that both σ_1 and σ_2 begin at the same state circle C, running downstream, and terminate in the same polyhedral region R. Then the following hold.*

(1) *There are disjoint right-down staircases for the σ_i, such that σ_1 runs downstream along each segment of the first staircase and σ_2 runs downstream along each segment of the second staircase.*
(2) *The terminal endpoint of each σ_i is adjacent to the last step (state circle) of its staircase.*
(3) *The j-th step of the first staircase is on the same state circle as the j-th step of the second staircase, except possibly the very last step.*
(4) *The arcs σ_1 and σ_2 cannot terminate on the same white face.*

Proof. Conclusions (1) and (2) will follow from Lemma 3.9 (Staircase extension), as soon as we verify that this lemma applies to the entire length of σ_1 and σ_2. That is, we need to check that each time σ_i enters a non-prime half-disk through a non-prime arc, it leaves that half-disk.

Suppose, for a contradiction, that σ_1 enters some non-prime half-disk through a non-prime arc, and does not leave it. All such half-disks are ordered by inclusion. Let R_1 be the *largest* such non-prime half-disk. Let α_1 be the non-prime arc through which σ_1 enters R_1, and let C_1 be the state circle to which it is attached. Since σ_2 also terminates inside $R \subset R_1$, and is disjoint from σ_1, it must cross into R_1 by crossing C_1.

Let γ denote the portion of σ_1 from C to α_1. By Lemma 3.9 (Staircase extension), there is a right-down staircase corresponding to γ. Thus C is connected to C_1 by a sequence of segments, and adjacent to the last such segment is a tentacle that meets the non-prime switch corresponding to α_1. Since the arc α_1 is next to the last stair, it is on the same side of C_1 as the stair. It follows that C and α_1 are on the same side of C_1. Thus σ_2 must actually cross C_1 twice, and by Lemma 3.11 (Utility lemma), it does so first running upstream, then running downstream.

But σ_2 left C running downstream. By Lemma 3.9 (Staircase extension), the only way σ_2 can later cross C_1 running upstream is if σ_2 crossed over a non-prime arc α_2 with endpoints on C_2, where α_2 separates C_1 from C. Let R_2 be the non-prime half-disk bounded by α_2 and C_2 and containing R_1. Since $R_1 \subset R_2$, σ_1 must also enter R_2, and it must do so by crossing C_2. Since σ_1 enters R_1 through non-prime arc α_1 (and not through a state circle), we conclude that $R_1 \neq R_2$.

Fig. 3.6 There exists a
closed curve in H_A of the
form of the *dotted line* above,
where the arc with *wider dots*
lies entirely in a region of the
complement of $H_A \cup (\bigcup \alpha_i)$

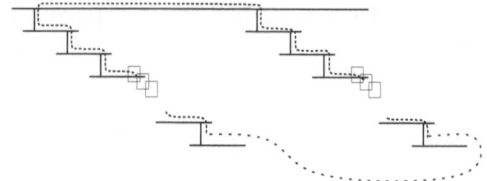

By applying to σ_1 the argument we used for σ_2 above, we conclude that σ_1 must cross C_2 twice, first running upstream and then downstream. Again, σ_1 cannot run upstream after leaving C in the downstream direction, unless R_2 is contained in a non-prime half-disk that σ_1 enters through a non-prime arc. But by construction, R_1 is the largest such half-disk, contradicting the strict inclusion $R_1 \subsetneq R_2$. This proves (1) and (2).

To prove (3), let $C = C_0, C_1, \ldots, C_m$ be the steps of the staircase of σ_1. Note that σ_1 runs downstream across each C_i (for $i = 0, \ldots, m - 1$). Thus, by Lemma 3.11 (Utility lemma), once σ_1 crosses a circle C_i, it may not cross it again. In other words, $C_0, C_1, \ldots, C_m$ are nested, and σ_1 runs deeper into this chain of nested circles.

Similarly, let $C = D_0, D_1, \ldots, D_n$ be the steps of the staircase of σ_2. Again, σ_2 runs downstream along $D_0, \ldots, D_{n-1}$, and cannot cross these circles a second time. Thus $D_0, \ldots, D_n$ are also nested.

By hypothesis, the terminal ends of σ_1 and σ_2 are in the same polyhedral region R. By the above work, each σ_i enters this region R by crossing a state circle running downstream. (Otherwise, σ_i would enter a non-prime half-disk across a non-prime arc without exiting, and we have ruled out this possibility.) Thus σ_1 enters R by crossing C_{m-1}, while σ_2 enters R by crossing D_{n-1}. Since the C_i are nested, as are the D_j, the only way this can happen is if $m = n$, and the stairs $C_j = D_j$ coincide for $j = 0, \ldots, n - 1$.

For (4), suppose that σ_1 and σ_2 terminate at the same white face W. Then we can draw an arc β entirely contained in W which meets the ends of both σ_1 and σ_2. Recall that a white face corresponds to a region of the complement of $H_A \cup (\bigcup \alpha_i)$. Thus the arc β corresponds to an arc, which we still denote β, in the complement of $H_A \cup (\bigcup \alpha_i)$ which meets the final segment of each right-down staircase on the right side of that segment, when the staircases are in right-down position. The two staircases, the state circle at the top, and the arc β form a loop in the sphere on which the graph H_A lies. See Fig. 3.6.

By conclusion (3), all steps of the staircases, except for the last, are on the same state circles. Note that the bottom stair C_n on the left is not inside the shown bounded region enclosed by the dotted curve β, but both ends of the bottom stair D_n on the right are inside the region enclosed by β. Since $C_j = D_j$ for $j = 0, \ldots, n - 1$, i.e. all stairs but the last connect from left to right, the two ends of the bottom right stair D_n must connect to each other only (and to none of the other state circles within the dotted curve), to form a state circle that does not intersect the dotted line at all, but lies entirely within it.

But then the arc β can be pushed to have both endpoints lying on the state circle C_{n-1} just above the bottom segment. It then gives a non-prime arc. By maximality of our polyhedral decomposition, Definition 2.19, there must be a collection of non-prime arcs $\alpha_{j_1}, \ldots, \alpha_{j_k}$ from our maximal decomposition so that the collection $\beta \cup (\cup \alpha_{j_i})$ bounds no state circles in its interior. But then one of these α_{j_i} must separate the bottom stair on the left from the bottom stair on the right. This non-prime arc would separate the bottom stairs into two distinct regions of the complement of $H_A \cup (\cup \alpha_i)$, contradicting our assumption that β lies in a single such region.   □

3.3 Bigons and Compression Disks

In an ideal polyhedral decomposition, any properly embedded essential surface (with or without boundary) can be placed into normal form. See, for example, Lackenby [58] or Futer and Guéritaud [30].

Definition 3.15. A surface in *normal form* satisfies five conditions:

(i)   Its intersection with ideal polyhedra is a collection of disks;
(ii) Each disk intersects a boundary edge of a polyhedron at most once;
(iii) The boundary of such a disk cannot enter and leave an ideal vertex through the same face of the polyhedron;
(iv) The surface intersects any face of the polyhedra in arcs, rather than simple closed curves;
(v)   No such arc can have endpoints in the same ideal vertex of a polyhedron, nor in a vertex and an adjacent edge.

Definition 3.16. A disk of intersection between a polyhedron and a normal surface is called a *normal disk*. For example, a *normal bigon* is a normal disk with two sides, which meets two distinct edges of its ambient polyhedron. Note that in a checkerboard colored polyhedron, one face met by a normal bigon must be white, and the other shaded.

Recall that, in Definition 2.20, we said that a polyhedron is prime if each pair of faces meet along at most one edge. This is equivalent to the absence of normal bigons.

Recall as well that our choice of a maximal collection of non-prime arcs may not have been unique, as pointed out just after Definition 2.19. However, using the idea of normal bigons, one can show that the prime polyhedral decomposition, obtained in Theorem 3.12, is unique. Because the result is not needed for our applications, we only outline the argument in the remark below. We point the reader to Atkinson [11] for more details.

Remark 3.17. One can see that the pieces of the prime decomposition are unique, as follows. We know, from Lemma 2.13, that the lower 3-balls are ideal polyhedra with 4-valent ideal vertices. For each lower polyhedron P, we may place a dihedral angle of $\pi/2$ on each edge, and construct an orbifold $\mathscr{O}_P$ by doubling P along its boundary. $\mathscr{O}_P$ is topologically the 3-sphere, with singular locus the planar 1-skeleton of P. Because we have doubled a dihedral angle of $\pi/2$, every edge in the singular locus has cone angle π.

There is a version of the prime decomposition for orbifolds, which involves cutting $\mathscr{O}_P$ along *orbifold spheres*, namely 2-dimensional orbifolds with positive Euler characteristic. Let S be one such orbifold sphere. In our setting, because the singular locus is a 4-valent graph, S must have an even number of cone points. Since the 1-skeleton of P is connected, the orbifold sphere S must intersect the singular locus, hence must have at least two cone points, with angle π. Therefore, since each singular edge has angle π, and S has positive Euler characteristic, it must have exactly two cone points.

Recall (e.g. from [11, 81]) that the prime decomposition of the orbifold $\mathscr{O}_P$ is equivariant with respect to the reflection along ∂P. Thus any orbifold sphere S is constructed by doubling a normal bigon in P. Since the prime decomposition of $\mathscr{O}_P$ is unique, and corresponds to cutting P along normal bigons, it follows that the decomposition of P along normal bigons is also unique.

The following proposition shows that our earlier definition of *prime decomposition* along non-prime arcs actually results in prime polyhedra. This, in turn, will be important in proving that the state surface S_A is essential in the link complement (Theorem 3.19).

Proposition 3.18 (No normal bigons). *Let $D(K)$ be an A-adequate link diagram, and let S_A be the all-A state surface of D. A prime decomposition of $S^3\backslash\backslash S_A$ into 3-balls, as in Definition 2.19, gives polyhedra which contain no normal bigons. In other words, every polyhedron is prime.*

Proof. Recall that by Lemma 2.21, part (5), the lower polyhedra are prime. Since a normal bigon is the obstruction to primeness, the lower polyhedra do not contain any normal bigons.

Suppose, by way of contradiction, that there exists a normal bigon in the upper polyhedron. Then its boundary consists of two arcs, one, γ_s embedded in a shaded face, and one, γ_w embedded on a single white disk W. Consider the arc γ_s in the shaded face. We may homotope this arc to lie on the spine of the shaded face. Since the spine is a tree, by Theorem 3.12, there is a unique embedded path between any pair of points on the tree. Hence γ_s is simple with respect to the shaded face.

First, note that γ_s must cross some state circle, for if not, γ_s remains on tentacles and non-prime switches adjacent to the same state circle C_0, and so γ_w contradicts part (ii) of the definition of normal, Definition 3.15.

So γ_s crosses a state circle C. The endpoints of γ_s are both on W, which means γ_s crosses C twice. If we cut out the middle part of γ_s (from C back to C), we obtain two disjoint sub-arcs from C to W. If we orient these sub-arcs away from C toward W, Lemma 3.11 (Utility lemma) implies they run downstream from C. Now, part (4) of Lemma 3.14 (Parallel stairs) says that the ends of γ_s cannot both be on W, which is a contradiction.   □

Recall that the state surface S_A may not be orientable. In this case, Definition 1.3 on p. 5 says that S_A is *essential* if the boundary $\widetilde{S_A}$ of its regular neighborhood is incompressible and boundary-incompressible. Since $S^3 \backslash\backslash \widetilde{S_A}$ is the disjoint union of $M_A = S^3 \backslash\backslash S_A$ and an I-bundle over S_A, the computation of the guts is not affected by replacing S_A with $\widetilde{S_A}$.

Theorem 3.19 (Ozawa). *Let D be a (connected) diagram of a link K. The surface S_A is essential in $S^3 \setminus K$ if and only if D is A-adequate.*

Proof. If D is not A-adequate, then there is an edge of H_A meeting the same state circle at each of its endpoints. To form S_A, we attach a twisted rectangle with opposite sides on a disk bounded by that same state circle. Note in this case, S_A will be non-orientable. The boundary of a disk E runs along S_A, over the twisted rectangle, meets the knot at the crossing of the rectangle, then continues along S_A through the disk bounded by that state circle. This disk E will give a boundary compression disk for $\widetilde{S_A}$, as follows. A regular neighborhood of S_A will meet E in a regular neighborhood of $\partial E \cap S_A$. Hence $E \setminus N(\partial E \cap S_A)$ is a compression disk for $\widetilde{S_A}$.

Now, suppose D is A-adequate, and let $\widetilde{S_A}$ be the boundary of a regular neighborhood of S_A. This orientable surface is the non-parabolic part of the boundary of M_A. If $\widetilde{S_A}$ is compressible, a compressing disk E has boundary on $\widetilde{S_A}$. Since $S^3 \backslash\backslash \widetilde{S_A}$ is the disjoint union of an I-bundle over S_A and M_A, the disk E must be contained either in the I-bundle or in M_A. It cannot be in the I-bundle, or in a neighborhood of $\widetilde{S_A}$ it would lift to a horizontal or vertical disk, contradicting the fact that it is a compression disk. Hence E lies in M_A.

Put the compressing disk E into normal form with respect to the polyhedral decomposition of M_A. The intersection of E with white faces contains no simple closed curves, so all intersections of E and the white faces are arcs. Consider an outermost arc. This has boundary a single arc on a white face, and a single arc on a shaded face. Hence it cuts off a normal bigon, which is a contradiction of Proposition 3.18 (No normal bigons). So the surface $\widetilde{S_A}$ is incompressible.

If $\widetilde{S_A}$ is boundary compressible, then a boundary compression disk E again lies in M_A rather than the I-bundle. Its boundary consists of two arcs, one on $\widetilde{S_A}$, which we denote β, and one which lies on the boundary of $S^3 \setminus K$ (the parabolic locus), which we denote α. Put E in normal form. First, we claim the arc α on $\partial(S^3 \setminus K)$ lies in a single polyhedron on a single ideal vertex. If not, it must meet one of the white faces of the polyhedron. Take an outermost arc of intersection of the white faces with E which cuts off a disk E' whose boundary contains a portion of the arc α. Either E' has an edge on a white face and an edge on α, in which case the surface E contradicts the first part of condition (v) of the definition of normal, or else E' has an edge on a white face, an edge on α, and an edge on S_A. In this case, E contradicts the second part of condition (v). Hence α lies entirely within one polyhedron.

Consider arcs of intersection of E with white faces. An outermost such arc must contain an ideal vertex, or we get a normal bigon as above, which is a contradiction. But if E' is outermost and E' contains an ideal vertex, then $E \setminus E'$ is a disk

which does not contain an ideal vertex. Again we get a contradiction looking at
the outermost arc of intersection of $E \setminus E'$ with white faces.   □

Lemma 3.20. *Every white face of the polyhedral decomposition is boundary
incompressible in M_A.*

Proof. If E is a boundary compression disk for a white face, it can be placed in
normal form. Then, as above, E must contain an outermost normal bigon, which
contradicts Proposition 3.18 (No normal bigons).   □

Recall that a link diagram D is *prime* if any simple closed curve which meets
the diagram transversely exactly twice does not bound crossings on each side.
Theorem 3.19 has the following corollary that shows that for prime, non-split links,
working with prime diagrams is not a restriction. Starting in Chap. 6, we will restrict
to prime adequate diagrams.

Corollary 3.21. *Suppose that K is an A-adequate, non-split, prime link. Then
every A-adequate diagram $D(K)$ without nugatory crossings is prime.*

Proof. Suppose $D(K)$ is an A-adequate diagram of K and let γ denote a simple
closed curve on the projection plane that intersects $D(K)$ at exactly two points.
Now γ splits $D(K)$ into a connect sum of diagrams $D_1 \# D_2$. Since K is prime,
one of them, say D_1, must be an A-adequate diagram of K, and D_2 must be an
A-adequate diagram of the unknot. The state surface S_A splits along an arc of γ
into surfaces S_1 and S_2, where S_i is the all-A state surface of D_i, $i = 1, 2$. By
Theorem 3.19, S_2 is incompressible, and thus it must be a disk. The graph $\mathbb{G}_A(D_2)$
is a spine for S_2. Since S_2 is a disk, $\mathbb{G}_A(D_2)$ is a tree. But then each edge of $\mathbb{G}_A(D_2)$
is separating, hence each crossing is nugatory. Since we assumed that D contains
no nugatory crossings, D_2 must be embedded on the projection plane. Thus $D(K)$
is prime, as desired.   □

The converse to Corollary 3.21 is open. See Problem 10.6 in Chap. 10.

3.4 Ideal Polyhedra for σ-Homogeneous Diagrams

In this section, we show that the decomposition for σ-homogeneous diagrams discussed in Sect. 2.4 becomes an ideal polyhedral decomposition under the additional
hypothesis of σ-adequacy. The arguments are almost identical to the already-
discussed case of A-adequate links. Thus our exposition here will be brief,
indicating only the cases where the argument calls for slight modifications.

In the σ-homogeneous setting, shaded faces decompose into portions associated
with a directed spine. An edge of the directed spine lies in each tentacle, and runs
adjacent to a segment and then along a state circle. The only difference now is that
when we are in a polyhedral region for which each resolution is the B-resolution,
these directed edges run left-down rather than right-down. Innermost disks are
still sources, and non-prime arcs give rise to switches (non-prime switches). The
resulting pieces are illustrated in Fig. 3.7, which should be compared to Fig. 3.2.

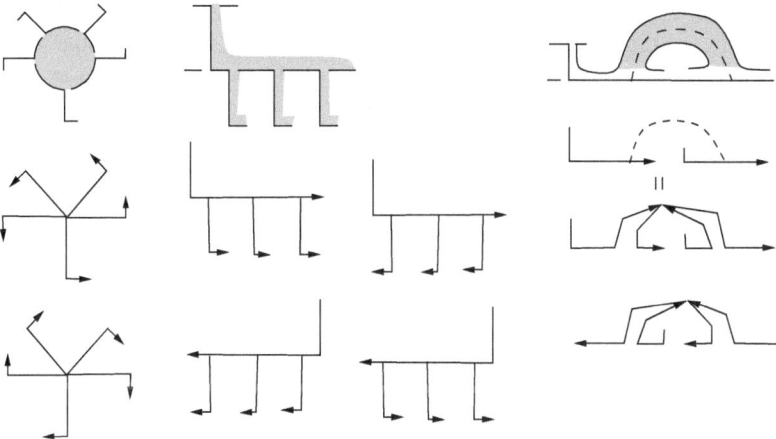

Fig. 3.7 Building blocks of the directed spine of a *shaded* face, in a σ-homogeneous diagram

As before, when an oriented arc in a shaded face runs in the direction of the directed spine, we say it is running *downstream*. Otherwise, it is running *upstream*. When such an arc has been homotoped to run monotonically through each tentacle, innermost disk, and non-prime switch, and to meet each at most once, we say it is *simple with respect to the shaded face*.

These definitions agree with Definitions 3.1 and 3.2, modified to accommodate left-down edges. Similarly, we have the following definition.

Definition 3.22. A *staircase* is an alternating sequence of state circles and segments. The direction of the staircase is determined by the directions of tentacles running along those staircases, which are determined by the resolution. Those of the A-resolution run "right-down". Those of the B-resolution run "left-down". All stairs in the same component of the complement of s_σ run in the same direction, by σ-homogeneity.

It turns out that the existence of a directed staircase is all that is needed for our main results. "Right-down-ness" and "left-down-ness" are only peripheral, and the theory developed so far in this chapter so far goes through without a problem. Hence we may prove the following analogue of Theorem 3.12.

Theorem 3.23. *Let σ be an adequate, homogeneous state of a diagram D. Then the decomposition described above gives a polyhedral decomposition of the surface complement M_σ into 4-valent ideal polyhedra.*

Proof. By σ-homogeneity, each lower polyhedron is identical to a polyhedron in Menasco's decomposition of an alternating link, which corresponds to the subgraph of H_σ coming from a polyhedral region. As for the upper polyhedron, ideal vertices are 4-valent, and white faces are simply connected. We need to show that shaded faces are simply connected in the σ-homogeneous case. Each shaded face deformation retracts to a directed spine, and we need to show this spine is a tree. The result follows from a sequence of lemmas established in the previous sections concerning how these directed graphs may be super-imposed on H_σ. The proofs of these lemmas work equally well when staircases run "right-down" and "left-down," as they will when A and B resolutions are mixed. What is key in all the proofs of these lemmas is that edges of the graph corresponding to the shaded faces have a direction, and the direction only changes in non-prime switches. In addition, the proofs repeatedly use the hypothesis that the state σ defining the graph H_σ is adequate (recall Example 3.5). Hence the following technical lemmas generalize without any modification of the proofs, except to remove the words "right-down" and replace "A-adequate" with "σ-adequate."

Lemma 3.4 (Escher stairs): No staircase forms a loop, and no staircase has its top and bottom on the same state circle.

Lemma 3.7 (Shortcut lemma): If a directed arc γ in a shaded face runs across a non-prime arc α with endpoints on a state circle C, and then upstream, the arc γ must exit the non-prime half-disk bounded by α and C by running downstream across C.

Lemma 3.9 (Staircase extension): If γ runs downstream across a state circle, and every time γ crosses a non-prime arc with endpoints on a state circle C, the arc γ exits the non-prime half-disk bounded by α and C, then γ defines a staircase such that γ is adjacent to each segment of the staircase, running downstream.

Lemma 3.10 (Downstream): For γ as above, it must cross the last state circle of the staircase running downstream.

Lemma 3.11 (Utility lemma): Let γ be a simple, directed arc in a shaded face, which starts and ends on the same state circle C. Then γ starts by running upstream from C, and then terminates at C while running downstream. Furthermore, γ cannot intersect C more than two times.

Now the proof of Theorem 3.12 goes through verbatim, only replacing H_A with H_σ. Hence the upper polyhedron is also a 4-valent ideal polyhedron.   □

Once we have a polyhedral decomposition of M_σ for a σ-adequate, σ-homogeneous diagram, we may use this to generalize Proposition 3.18 and Theorem 3.19 in the setting of σ-adequate and σ-homogeneous diagrams.

In order to do so, we need Lemma 3.14 (Parallel stairs). More specifically, we need part (4) of that lemma, but we state the entire lemma for completeness.

Lemma 3.14 (Parallel stairs): Let σ_1 and σ_2 be simple, disjoint, directed arcs through the spines of shaded faces F_1 and F_2. (These shaded faces are allowed to coincide, so long as the σ_i are disjoint.) Suppose that both σ_1 and σ_2 begin at the same state circle C, running downstream, and terminate in the same polyhedral region R. Then

(1) There are disjoint staircases for the σ_i, such that σ_1 runs downstream along each segment of the first staircase and σ_2 runs downstream along each segment of the second staircase.

(2) The terminal endpoint of each σ_i is adjacent to the last step (state circle) of its staircase.
(3) The j-th step of the first staircase is on the same state circle as the j-th step of the second staircase, except possibly the very last step.
(4) The arcs σ_1 and σ_2 cannot terminate on the same white face.

As in the case of the A-adequate links, the proof constructs staircases for σ_1 and σ_2, using Lemma 3.9 (Staircase extension). Furthermore, the proof of the (generalized) lemma requires σ-homogeneity, in that if both arcs running downstream along the staircases end in tentacles meeting the same white face, then at the bottom the arcs are both either running in the right-down or the left-down direction, and we obtain a diagram as in Fig. 3.6 or its reflection. That is, we obtain a sequence of stairs on the right and the left, with bottom segments of the stairs connected by an arc β in the complement of $H_\sigma \cup (\bigcup \alpha_i)$ which runs from the right side of one last segment to the right side of the other, or from the left side of one last segment to the left side of the other. In either case, the argument of the proof of that lemma will still imply that stairs connect left to right, excepting the two bottom stairs, and that the arc β can have its endpoints pushed to the state circle just above both bottom stairs to give a non-prime arc, contradicting maximality of our choice of a system of non-prime arcs. Then the proof of Proposition 3.18 goes through verbatim to give the following.

Proposition 3.24 (No Normal Bigons). *Let $D(K)$ be a link diagram with an adequate, homogeneous state σ, and let S_σ be the state surface of σ. Then the decomposition of $S^3 \backslash\backslash S_\sigma$ as above gives polyhedra without normal bigons. In other words, every polyhedron is prime.*   □

Finally, given these pieces, we obtain Theorem 3.19 in this setting, without modification to the proof. The theorem is originally due to Ozawa [76].

Theorem 3.25 (Ozawa). *Let D be a (connected) diagram of a link K, such that D is σ-homogeneous for some state σ. The surface S_σ is essential in $S^3 \setminus K$ if and only if D is σ-adequate.*   □

Chapter 4
I-Bundles and Essential Product Disks

Recall that we are trying to relate geometric and topological aspects of the knot complement $S^3 \setminus K$ to quantum invariants and diagrammatic properties. So far, we have identified an essential state surface S_A, and we have found a polyhedral decomposition of $M_A = S^3 \backslash\backslash S_A$. On the one hand, the surface S_A is known to have relations to the Jones and colored Jones polynomials [21, 23, 36]. On the other hand, the Euler characteristic of the *guts* of M_A, whose definition is recalled immediately below, is known to have relations to the volume [6]. As mentioned in the introduction, we will see in Chap. 9 that the Euler characteristic of the guts of M_A forms a bridge between geometric and quantum invariants. In this chapter, we take a first step toward computing this Euler characteristic, using the polyhedral decomposition from Chap. 3.

By the annulus version of the JSJ decomposition, there is a canonical way to decompose $M_A = S^3 \backslash\backslash S_A$ along a collection of essential annuli that are disjoint from the parabolic locus. (In our case, recall from Definition 1.2 that the parabolic locus of M_A will be the remnant of the boundary of a regular neighborhood of K in M_A.) The JSJ decomposition yields two kinds of pieces: the *characteristic submanifold*, consisting of I-bundles and Seifert fibered pieces, and the *guts*, which admit a hyperbolic metric with totally geodesic boundary. We consider the components of the characteristic submanifold of M_A which affect Euler characteristic. In this chapter, we show that such components decompose into well-behaved pieces. In particular, we show that they are spanned by essential product disks (Definition 4.2) which are each embedded in a single polyhedron of the polyhedral decomposition of M_A from Chap. 3. This is the content of Theorem 4.4, which is the main result of the chapter.

4.1 Maximal *I*-Bundles

Let B be a component of the characteristic submanifold of M_A; so B is either a Seifert fibered component or an I-bundle. Our first observation implies that only I-bundles can have non-trivial Euler characteristic.

D. Futer et al., *Guts of Surfaces and the Colored Jones Polynomial*, Lecture Notes in Mathematics 2069, DOI 10.1007/978-3-642-33302-6_4,
© Springer-Verlag Berlin Heidelberg 2013

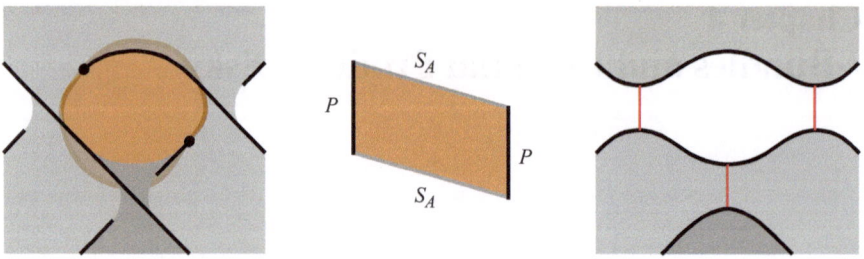

Fig. 4.1 *Left*: an EPD in M_A. Center: the product structure of the EPD. *Right*: the corresponding subgraph of H_A contains a 2-edge loop. In this example, the two segments come from different twist regions in the link diagram

Lemma 4.1. *Let B be a component of the characteristic submanifold of M_A. Then $\chi(B) \leq 0$, and B can come in one of two flavors:*

(1) *If $\chi(B) < 0$, then B is an I-bundle. We call such components* non-trivial.
(2) *If $\chi(B) = 0$, then B is a solid torus. We call these solid tori* trivial.

The reason for this terminology is that removing a solid torus B does not affect the Euler characteristic of what remains. Thus, for computing the Euler characteristic of the guts, one only needs to worry non-trivial I-bundles.

Proof. Recall that by Lemma 2.4, M_A is topologically a handlebody, with non-positive Euler characteristic. To find the characteristic submanifold, we cut M_A along essential annuli. Thus every component of the complement of these annuli will again have non-positive Euler characteristic.

The component B is either an I-bundle or a Seifert fibered piece. If B is an I-bundle, then $\chi(B)$ will vanish if and only if B is a solid torus (viewed as an I-bundle over an annulus or Möbius band).

Next, suppose B is a Seifert fibered component. Since M_A is a handlebody, its fundamental group is free, and cannot contain a $\mathbb{Z} \times \mathbb{Z}$ subgroup. On the other hand, B is a Seifert fibered 3-manifold with boundary. Its base orbifold O must be an orbifold with boundary. But B cannot contain any essential tori, hence the orbifold O cannot contain any essential loops. This is possible only if O a disk with at most one cone point, and B is a solid torus, with zero Euler characteristic.   □

The main result of this chapter is that all the non-trivial components of I-bundle can be found by studying *essential product disks*.

Definition 4.2. An *essential product disk* (EPD for short) is a properly embedded essential disk in M_A, whose boundary meets the parabolic locus of M_A twice. See Fig. 4.1.

Essential product disks play an important role in computing guts in Lackenby's volume estimates for alternating links [58]. They will play an important role in our setting as well.

Recall that M_A is a handlebody, so certainly it admits a number of compression disks. However, a compression disk for M_A that is disjoint from the parabolic locus would be a compression disk for S_A; by Theorem 3.19, such disks cannot exist. Similarly, a compression disk for M_A that meets the parabolic locus only once would be a boundary compression disk for S_A; by Theorem 3.19, such disks also cannot exist. Thus essential product disks can be seen as the simplest compression disks for M_A.

Notice that a regular neighborhood of an essential product disk is an I-bundle, and is thus contained in the characteristic submanifold of M_A.

Definition 4.3. Let B be an I-bundle in the characteristic submanifold of M_A. We say that a finite collection of disjoint essential product disks $\{D_1, \ldots, D_n\}$ *spans* B if $B \setminus (D_1 \cup \cdots \cup D_n)$ is a finite collection of prisms (which are I-bundles over a polygon) and solid tori (which are I-bundles over an annulus or Möbius band).

Our main result in this chapter is the following theorem, which reduces the problem of understanding the I-bundle in the characteristic submanifold of M_A to the problem of understanding and counting EPDs in individual polyhedra. For instance, the EPD of Fig. 4.1 is embedded in a lower polyhedron. In the following chapters we will study such EPDs.

Theorem 4.4. *Let B be a non-trivial component of the characteristic submanifold of M_A. Then B is spanned by a collection of essential product disks $D_1, \ldots, D_n$, with the property that each D_i is embedded in a single polyhedron in the polyhedral decomposition of M_A.*

The proof of the theorem will occupy the remainder of this chapter. Before we give an outline of the proof, we need the following definition.

Definition 4.5. A surface S in M_A is *parabolically compressible* if there is an embedded disk E in M_A such that:

(i) $E \cap S$ is a single arc in ∂E;
(ii) The rest of ∂E is an arc in ∂M_A that has endpoints disjoint from the parabolic locus P of M_A and that intersects P in at most one transverse arc;
(iii) E is not parallel into S under an isotopy that keeps $E \cap S$ fixed and keeps $E \cap P$ on the parabolic locus.

We say E is a *parabolic compression disk*. (See [58, Fig. 5].)

Definition 4.5 differs slightly from the corresponding definition in Lackenby's work [58, Page 209]. Conditions (i) and (ii) are exactly the same, while our condition (iii) is slightly less restrictive.

If D is an essential product disk and E is a parabolic compression for D, then compressing D to the parabolic locus along E will produce a pair of new essential product disks, D' and D''. See Fig. 4.2. Observe that if $D, D_1, \ldots, D_n$ span an I-bundle B, then $D', D'', D_1, \ldots, D_n$ will span B as well. Thus we may perform parabolic compressions at will, without losing the property that the disks in question span B.

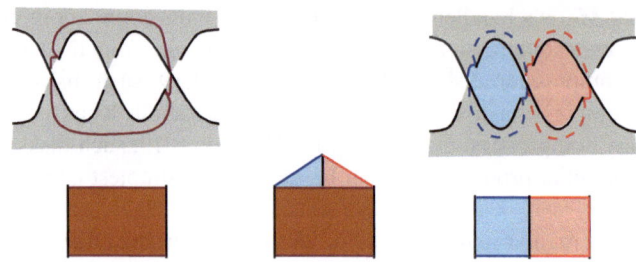

Fig. 4.2 The EPD shown in the *left panel* parabolically compresses to the EPDs shown in the *right panel*

Proof (Top-level proof of Theorem 4.4). The argument has three main steps:

1. Given a non-trivial component B of the characteristic submanifold of M_A we show B meets the parabolic locus (Proposition 4.18).
2. We show that if a component B as above meets the parabolic locus, it is spanned by essential product disks (Proposition 4.19).
3. We show that every essential product disk in M_A parabolically compresses to a collection of essential product disks, each of which is embedded in a single polyhedron (Proposition 4.21).

Step 2 will be completed by straightforward topological argument. On the other hand, Steps 1 and 3 require a number of technical tools that we will develop in the next sections. Thus we postpone the proofs of all three propositions until the end of the chapter. Modulo these propositions, the proof of Theorem 4.4 is complete. □

Here is the outline of the rest of the chapter. Section 4.2 uses normal surface theory to examine pieces of the boundaries of I-bundles. Section 4.3 uses tentacle chasing arguments to force parabolic compressions. In Sect. 4.4, we put it all together to finish the proof of Steps 1–3. In Sect. 4.5, we discuss the (straightforward) extension to σ-adequate, σ-homogeneous diagrams.

4.2 Normal Squares and Gluings

In this section, we will consider properties of normal squares (i.e. normal disks with four sides; see Definitions 3.15 and 3.16). In Lemma 4.6, we will see that normal squares arise naturally as intersections of annuli, boundary components of our characteristic submanifold, and our ideal polyhedra. With this in mind, we need to examine how normal squares glue across white faces of the ideal polyhedra. The results of this section, which are of a somewhat technical nature, will be used to examine this gluing.

Lemma 4.6. *Let B_0 be a component of the maximal I-bundle for M_A with negative Euler characteristic. Then B_0 contains a product bundle $Y = Q \times I$, where Q is a* *pair of pants. Moreover, when put into normal form in a prime decomposition of M_A, the three annuli of ∂Y are composed of disjointly embedded normal squares.*

The proof of Lemma 4.6 should be compared to that of [3, Lemma 7.1].

Proof. Since B_0 is a 3-dimensional submanifold of S^3, it must be orientable. Thus B_0 is either the product I-bundle over an orientable surface F, or the twisted I-bundle over a non-orientable surface F. In either case, since $\chi(B_0) = \chi(F) < 0$, F contains a pair of pants Q. (In a non-orientable surface, cutting a once-punctured Möbius band along an orientation-reversing closed curve produces a pair of pants.) The I-bundle over the pair of pants Q must be trivial, so B_0 contains a product bundle $Y = Q \times I$.

Consider the three annuli of ∂Y. We view the union of these three annuli as a single embedded surface. Move this surface into normal form in the polyhedral decomposition of M_A, keeping the surface embedded. The annuli of ∂Y stay disjoint. The intersection of the annuli with the faces of the polyhedra cuts the annuli into polygons, each of which must have an even number of edges due to the checkerboard coloring of the polyhedra.

Consider an arc of intersection between the white faces and an annulus $A \subset \partial Y$. If this arc α starts and ends on the same boundary circle of A, then A cuts off a bigon disk. An outermost such arc would cut off a normal bigon in a single polyhedron—but by Proposition 3.18, there are no normal bigons. Thus the arc α must run from one boundary circle of A to the other boundary circle. Because every arc of intersection between ∂Y and the white faces is of this form, every normal disk must be a square. □

While studying the checkerboard surfaces of alternating links, Lackenby has obtained useful results by super-imposing normal squares in the upper polyhedron onto normal squares in the lower polyhedron [58]. For alternating knots and links, the 1-skeleton of each polyhedron is the 4-valent graph of the link projection; thus there is a natural "identity map" from one polyhedron to the other. Lackenby's method will also be useful for our results, although we need to take some care defining maps between the upper and lower polyhedra.

For each white face W, the disk W appears as a face of the upper polyhedron and exactly one lower polyhedron. These two faces are glued via the gluing map, which is just the reverse of the cutting moves we did in Chap. 2 to form the polyhedra.

Definition 4.7. Let W be a white face of the upper polyhedron P, and suppose that W has n sides. For the purpose of defining continuous functions, picture W as a regular n-gon in $\mathbb{R}^2$. Let W' be the face of a lower polyhedron that is glued to W in the polyhedral decomposition. Then we define a *clockwise map* $\phi: W \to W'$ to be the composition of the gluing map with a $2\pi/n$ clockwise rotation. In other words, both the gluing map and the clockwise map send W to W', but the two maps differ by one side of the polygon.

Combinatorially, in the upper polyhedron, white faces are sketched with edges on tentacles and non-prime switches, and with vertices adjacent to a state circle at the

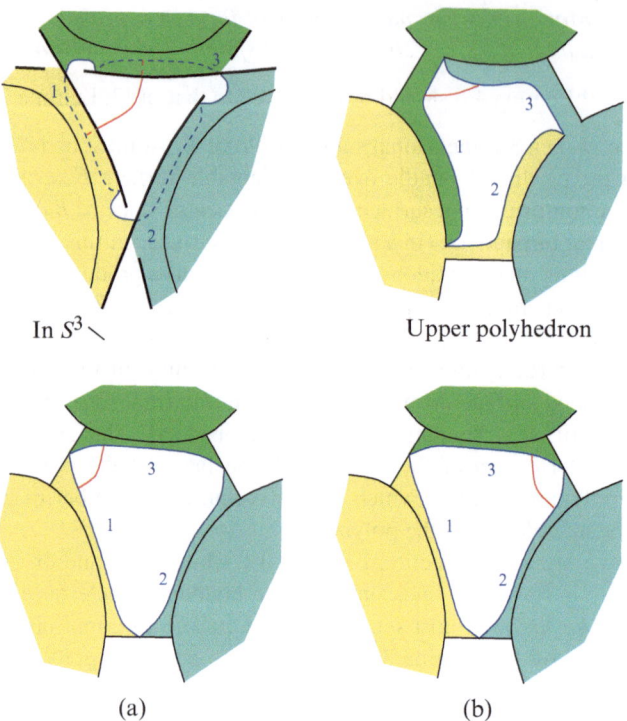

In $S^3 \diagdown$ Upper polyhedron

(a)   (b)

Fig. 4.3 An example showing the image of an arc β under (**a**) the gluing map, and (**b**) the clockwise map

top-right (bottom-left) at a crossing, or segment of H_A, as in the right of Fig. 2.10 on p. 28. However in the lower polyhedron, white faces are drawn with vertices in the center of segments of H_A, as in Fig. 2.9 on p. 28. The gluing map gives the white faces on the upper polyhedron a slight rotation counterclockwise, moving a vertex adjacent to a segment of H_A to lie at the center of that same segment, and then maps the region on the upper polyhedron to the corresponding region on the lower polyhedron by the identity. See Fig. 4.3. On the other hand, instead of rotating counterclockwise in the upper polyhedron to put vertices at the centers of segments of H_A, ϕ rotates clockwise to the nearest adjacent edge in the clockwise direction.

It is instructive to compare our setting with Menasco's polyhedral decomposition of alternating links [64]. In an alternating diagram $D(K)$, the state surface S_A is a (shaded) checkerboard surface for K, and the union of all the white faces of the polyhedra is the other (white) checkerboard surface S_B. If the 1-skeleta of both the top polyhedron and the bottom polyhedron are identified with the 4-valent graph of $D(K)$, then the gluing map rotates all white faces counterclockwise and all shaded faces clockwise. In other words, on all the white faces, the identity map differs from the gluing map by a clockwise rotation. Furthermore, the identity map is of course defined on the entire polyhedron, not just on the white faces.

In our case, the clockwise map ϕ is an analogue of the identity map, and also differs from the gluing map by a $2\pi/n$ clockwise rotation. In keeping with the analogy, the domain of definition of ϕ can be extended beyond the white faces (although not all the way to the entire top polyhedron).

Lemma 4.8. *Let U be a polyhedral region of the projection plane, that is, a region of the complement of $s_A \cup (\cup_i \alpha_i)$. Let $W_1, \ldots, W_n$ be the white faces in U, and let P' be the lower polyhedron associated to U. Then the clockwise map $\phi : W_1 \cup \ldots \cup W_n \to P'$ has the following properties:*

(1) *If x and y are points on the boundary of white faces in U that belong to the same shaded face of the upper polyhedron P, then $\phi(x)$ and $\phi(y)$ belong to the same shaded face of P'.*

(2) *Let $S \subset P$ be a normal square in the upper polyhedron, such that the white faces V, W intersected by S belong to U. Let $\beta_V = S \cap V$ and $\beta_W = S \cap W$. Then the arcs $\phi(\beta_V)$ and $\phi(\beta_W)$ can be joined along shaded faces to give a normal square $S' \subset P'$, defined uniquely up to normal isotopy. We write $S' = \phi(S)$.*

(3) *If S_1 and S_2 are disjoint normal squares in P, all of whose white faces belong to U, then S_1' and S_2' are disjoint normal squares in P'.*

Proof. For conclusion (1), let F be a shaded face of the upper polyhedron P, and let x and y be points on $(\partial F) \cap U$. Then x and y can be connected by an arc γ running through F, and we can make γ simple with respect to F (Definition 3.2). If the arc γ is parallel to an ideal edge e of the upper polyhedron, then $x, y \in e$, hence $\phi(x), \phi(y) \in \phi(e)$, and the conclusion holds. Otherwise, the arc γ must cross some state circle C, hence is non-trivial. Because both of its endpoints are in the same polyhedral region, in fact γ must cross C twice, first running upstream and then downstream by Lemma 3.11. Thus we can split γ into two disjoint arcs beginning at C, running downstream, and terminating in the same polyhedral region. By Lemma 3.14 (Parallel stairs), γ must run up and down a pair of right-down staircases, and by part (3) of that lemma, the first state circle C_1 that γ crosses when running from x to y must be the same as the last state circle C_1. Now $C_1 \subset \partial U$ corresponds to a shaded face F' of the lower polyhedron P' (see Fig. 2.9 on page 28). Thus both $\phi(x)$ and $\phi(y)$ must lie on the boundary of F'.

For (2), let $S \subset P$ be a normal square, such that the white faces V, W intersected by S belong to U. Let u, x, y, z be the four points of intersection between S and the edges of P, such that $u, x \in \partial V$ and $y, z \in \partial W$. Then x, y lie on the boundary of the same shaded face F. By conclusion (1), $\phi(x), \phi(y)$ belong to the same shaded face $F' \subset P'$. Since F' is simply connected by Theorem 3.12, $\phi(x)$ and $\phi(y)$ can be connected by a unique isotopy class of arc in F'. Similarly, $\phi(z)$ and $\phi(u)$ can be connected by a unique isotopy class of arc in a shaded face of P'. These normal arcs in shaded faces combine with the arcs $\phi(\beta_V) \subset \phi(V)$ and $\phi(\beta_W) \subset \phi(W)$ to form a normal square $S' \subset P'$, which is unique up to normal isotopy.

For conclusion (3), let S_1 and S_2 be disjoint normal squares in P, all of whose white faces belong to U. The arcs of S_1 and S_2 that lie in white faces of U are mapped homeomorphically (hence disjointly) to white faces in P'. Thus it remains to check that the arcs of S_1' and S_2' are also disjoint in the shaded faces. Suppose that both S_1 and S_2 pass through a shaded face F, disjointly. Then we can label points w, x, y, z, in clockwise order around ∂F, such that S_1 intersects ∂F at points w, x and S_2 intersects F at points y, z. Then the four points $\phi(w), \phi(x), \phi(y), \phi(z)$ are also arranged in clockwise order around a shaded face F' of P'. Thus S_1' and S_2' are disjoint in F'. $\square$

One part of proving the main result in this chapter is to show that certain normal squares in the upper polyhedron are parabolically compressible. For that, we will map them to squares in the lower polyhedra, using the clockwise map and Lemma 4.8, and consider their intersections with certain normal squares in the lower polyhedra. We use the following lemma, which is due to Lackenby [58, Lemma 7]. We include the proof for completeness.

Lemma 4.9. *Let* P *be a prime, checkerboard-colored polyhedron with 4-valent vertices. Let* S *and* T *be normal squares in* P, *which have been moved by normal isotopy into a position that minimizes their intersection number. Then* S *and* T *are either disjoint, or they have an essential intersection in two faces of the same color.*

Recall, from Definition 2.20 on p. 29, that a polyhedron P is prime if it contains no normal bigons. By Proposition 3.18 on p. 46, our polyhedra are all prime.

Proof. The four sides of S run through four distinct faces of the polyhedron, as do the four sides of T. A side of S intersects a side of T at most once. If all four sides of S intersect sides of T, then S and T are isotopic and can be isotoped off each other. So S and T intersect at most three times. However, ∂S and ∂T form closed curves on the boundary of the polyhedron, hence S intersects T an even number of times, so 0 or 2 times. If twice, suppose S and T intersect in faces of opposite color. Then each arc of $S \setminus T$ and $T \setminus S$ intersects the edges of the polyhedron an odd number of times. Hence one of the four complementary regions of $S \cup T$ has two points of intersection with the edges of the polyhedron in its boundary. Because the polyhedron is prime, this gives a bigon which cannot be normal, hence S and T can be isotoped off each other. $\square$

This lemma has the following useful consequence, illustrated in Fig. 4.4.

Lemma 4.10. *Let* P *be a prime, checkerboard-colored polyhedron with 4-valent vertices. Let* S *and* T *be normal squares in* P, *moved by normal isotopy to minimize their intersection number. Suppose that* S *and* T *pass through the same white face* W, *and that the edges* $S \cap W$ *and* $T \cap W$ *differ by a single rotation of* W *(clockwise or counterclockwise). Then exactly one of the following two conclusions holds:*

(1) *Each of* S *and* T *cuts off a single ideal vertex in* W, *and* S *and* T *are disjoint.*
(2) *Neither* S *nor* T *cuts off a single vertex in* W. *The two normal squares intersect in* W *and in another white face* W'.

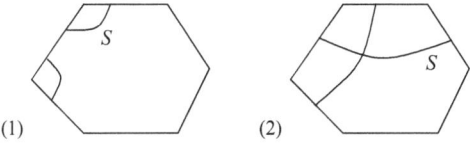

Fig. 4.4 Cases (1) and (2) of Lemma 4.10 are illustrated for an example. Note if (2) happens, S and T intersect in two *white* faces

Proof. First, suppose that S cuts off a single ideal vertex in W. Then so does T. Hence S and T do not intersect in W. By Lemma 4.9, we can conclude that if S and T intersect at all, they intersect in two shaded faces, F and G. Since W meets each shaded face at only one edge, $S \cap W$ and $T \cap W$ must run in parallel through W. Thus their intersections in F and G can be isotoped away, and S and T are disjoint. This proves (1).

Next, suppose that S does not cut off a single ideal vertex in W. Then neither does T. Thus, since $S \cap W$ and $T \cap W$ differ by a clockwise or counterclockwise rotation of W, they must have an essential intersection. By Lemma 4.9, they must also intersect in another white face W'. This proves (2). $\square$

When case (1) of Lemma 4.10 holds, note that there is a parabolic compression of S, through W, to the single ideal vertex of W that it cuts off. Similarly for T.

Definition 4.11. Let P be a truncated, checkerboard-colored ideal polyhedron. Then a *normal trapezoid* in P is a normal disk that passes through two shaded faces, one white face, and one truncated ideal vertex.

Trapezoids give the following analogue of Lemma 4.10.

Lemma 4.12. *Let P be a prime, checkerboard-colored polyhedron with 4-valent vertices. Let S be a normal square in P, and T a normal trapezoid. Suppose that S and T pass through the same white face W, and that their arcs of intersections with W differ by a single rotation (clockwise or counterclockwise). Then S and T are disjoint, and each of S and T is parabolically compressible to an ideal vertex of W.*

Proof. Let T' be a normal square obtained by pulling T off an ideal vertex, into a white face W'. Then S and T' are normal squares that satisfy the hypotheses of Lemma 4.10. Because T' cuts off a single ideal vertex of W', S and T' cannot intersect in that white face. Thus conclusion (1) of Lemma 4.10 holds: S and T' are disjoint, and each one cuts off a single ideal vertex in W. Thus S and T are also disjoint, and each one is parabolically compressible to an ideal vertex of W. $\square$

By applying Lemmas 4.10 and 4.12, we will show that many normal squares are also parabolically compressible. See e.g. the proof of Lemma 4.20 for a preview of the argument.

4.3   Parabolically Compressing Normal Squares

Results of Sect. 4.2 are enough to handle normal squares with sides in the same polyhedral region. Note this is all that occurs for alternating knots, as in [58]. In this section, we will use tentacle chasing arguments to extend our tools, so that we can deal with normal squares with their sides in different polyhedral regions. This is the content of the next proposition, which is the main result in this section.

Proposition 4.13. *Let S be a normal square in the upper polyhedron, with boundary consisting of arcs β_V, β_W on white faces V and W, and arcs σ_1, σ_2 on shaded faces. Suppose that V and W are in different polyhedral regions. Finally, suppose that S is glued to a normal square T at W. Then S cuts off a single ideal vertex in W, hence is parabolically compressible at W.*

Proposition 4.13 is a crucial ingredient for the proof of the main result of the chapter, Theorem 4.4, which is given in the next section. Before we give the proof of the proposition, we need to establish some technical lemmas. We advise the reader that only the statement of Proposition 4.13, and not those of the intermediate technical lemmas, are required for the proof of Theorem 4.4. Thus readers eager to get to the proof of the main result of the chapter may, at this point, move to the next section, on p. 67, without loss of continuity. However, several of the technical lemmas in the remainder of this section are repeatedly used in Chap. 6.

Lemma 4.14 (Opposite sides). *Let S be a normal square with boundary arcs β_V and β_W on white faces V and W, and arcs σ_1 and σ_2 on distinct shaded faces. Suppose σ_1 and σ_2 intersect the same state circle C. Then the intersections are in tentacles attached to edges on opposite sides of C, and C must separate V and W.*

Recall again that arcs in a shaded face can only intersect state circles at the heads of tentacles, adjacent to segments of H_A. (See Definitions 2.16 and 2.17, as well as Fig. 2.7, on p. 25.) Lemma 4.14 (Opposite sides) asserts that under the given hypotheses, σ_1 and σ_2 run adjacent to heads of tentacles attached to C, but adjacent to segments on opposite sides of C.

Proof. We will first show that C must separate W and V, and then that when we direct σ_1 and σ_2 to run across C away from V and toward W, one of σ_1, σ_2 runs upstream and one runs downstream. This will imply the result.

Suppose, by way of contradiction, that C does not separate V and W, but that both lie on the same side of C. Then both σ_1 and σ_2 must intersect C twice. Direct σ_1 and σ_2 away from V. We may assume each is simple with respect to its shaded face. Now Lemma 3.11 (Utility lemma) implies that both arcs cross C first running upstream, then running downstream. Consider the portion of the arcs running downstream. These are both running downstream from C, connected at their ends by β_W. This contradicts part (4) of Lemma 3.14 (Parallel stairs).

Now, suppose C separates V and W, but σ_1 and σ_2 run in the same direction across C. Then, switching V and W if necessary, we may assume that both σ_1 and σ_2 run away from V across C in the downstream direction. Again we have arcs

σ_1 and σ_2 running downstream from C, connected at their ends by β_W. Again this contradicts Lemma 3.14 (Parallel stairs).

Lemma 4.15 (Entering polyhedral region). *Let S be a normal square with boundary consisting of arcs β_V and β_W on white faces V and W, and arcs σ_1 and σ_2 on shaded faces. Suppose also that V and W are in distinct polyhedral regions R_V and R_W. Then (up to relabeling), when σ_1 and σ_2 are directed away from V towards W, we have the following:*

(1) *The arc σ_1 first enters R_W through a state circle C running in the downstream direction, and immediately connects to β_W (i.e., without intersecting any additional state circles or non-prime arcs).*

(2) *The arc σ_2 first enters R_W either through C running upstream, or through a non-prime arc with both endpoints on C. In any case, if σ_2 crosses C, then it must do so only once, running upstream.*

Proof. Since V and W are in distinct polyhedral regions, they are either on opposite sides of some state circle, or if they are not on opposite sides of any state circle, they are separated by a non-prime arc α with both endpoints on a state circle C. In the latter case, C does not separate V and W, nor does any state circle contained inside the non-prime half-disk bounded by α and C separate V and W. We distinguish two cases.

Case 1: Suppose that V and W are separated by a non-prime arc α with both endpoints on some state circle C, but that C does not separate V and W. Suppose also that within the non-prime half-disk bounded by α and C that contains W, no other state circle separates V and W. Without loss of generality, we may suppose that α is innermost with this property with respect to W, that is, that α is the non-prime arc with this property closest to W.

Notice that one of σ_1, σ_2 must cross C, since the arcs are on distinct shaded faces. After relabeling, we may assume σ_1 crosses C. Since V and W are on the same side of C, σ_1 must actually cross C twice. But then Lemma 3.11 (Utility lemma) implies that it crosses C first running upstream, then running downstream. So σ_1 crosses C running downstream when it enters R_W.

Since σ_1 is running downstream, it will be adjacent to some state circle C_4 attached to C by a segment of the graph H_A. By assumption, C_4 does not separate V and W.

Since σ_1 is going downstream, Lemma 3.10 (Downstream continues down) implies that it can only continue downstream next, or cross a non-prime arc with W inside, or exit the shaded face immediately to β_W. The first of these three possibilities cannot happen: σ_1 cannot continue downstream, else it crosses into C_4, and must cross back out, which is impossible by Lemma 3.11 (Utility lemma). The second possibility also cannot hold by assumption: α was assumed to be innermost with respect to W. Thus the only possibility is that σ_1 exits the shaded face immediately to β_W.

If σ_2 also crosses C, then it must do so twice, since its endpoints are not separated by C. Lemma 3.11 (Utility lemma) implies that σ_2 first crosses C running upstream, then crosses running downstream. But now σ_1 and σ_2 both cross C running downstream, then have endpoints attached at β_W. This contradicts Lemma 3.14 (Parallel stairs). So σ_2 does not cross C, which implies it enters W across a non-prime arc, as desired.

Case 2: Suppose that V and W are on opposite sides of some state circle. Then there is some such state circle which is closest to W. Call this state circle C_W. The arcs σ_1 and σ_2 must both intersect C_W. By Lemma 4.14 (Opposite sides), the intersections are in opposite directions, when σ_1 and σ_2 are both directed toward W. Relabel, if necessary, so that σ_1 is the arc running downstream toward W across C_W.

Since σ_1 is running downstream, it will be adjacent to some state circle C_4 attached to C_W by a segment of the graph H_A. Again since σ_1 is running downstream, Lemma 3.10 (Downstream continues down) implies that it can only go downstream next, or cross a non-prime arc with W inside, or exit the shaded face to β_W. The first possibility cannot happen: σ_1 cannot continue downstream, else it crosses into C_4, and must cross back out, which is impossible by Lemma 3.11 (Utility lemma). Suppose the second possibility holds, that is that σ_1 crosses a non-prime arc α with W inside. This non-prime arc α has both endpoints on C_4, and C_4 does not separate V and W by choice of C_W. Moreover, within the non-prime half-disk bounded by α and C_4 which contains W, no state circle can separate V and W, again by choice of C_W. Thus if this second possibility holds, we are in Case 1, with C_4 playing the role of C, and the lemma is true (after relabeling σ_1 and σ_2 again).

The only remaining possibility is that σ_1 exits the shaded face immediately to β_W after crossing C_W. This proves statement (1) of the Lemma, with $C = C_W$.

Finally, the fact that σ_2 must cross C running upstream, if it crosses C at all, follows immediately from the fact that σ_1 crosses C running downstream, and Lemma 4.14 (Opposite sides). $\qquad\square$

Lemma 4.16. *Let S be a normal square with boundary arcs β_V and β_W on white faces V and W, and arcs σ_1 and σ_2 on shaded faces. Suppose V and W are in distinct polyhedral regions R_V and R_W. Let C_W be the state circle in the conclusion of Lemma 4.15 (Entering polyhedral region), and relabel if necessary so that σ_1 and σ_2 are as in the conclusion of that Lemma, when directed away from V towards W. Suppose, moreover, that arc σ_2 does not immediately connect to β_W after it first enters R_W. Then, σ_2 runs across a state circle C_2 running upstream, then eventually crosses C_2 again into R_W running downstream, at which point it immediately connects to β_W (i.e. without crossing any other state circles).*

Note that Lemmas 4.15 and 4.16 imply that if σ_2 does not immediately connect to β_W, the region R_W is of the form shown in Fig. 4.5.

Essentially, what these two lemmas say is that σ_1 only enters the region R_W once, to connect to β_W. The arc σ_2, on the other hand, may enter R_W, then leave and travel elsewhere, but when it returns it will not leave again, but connect immediately to β_W.

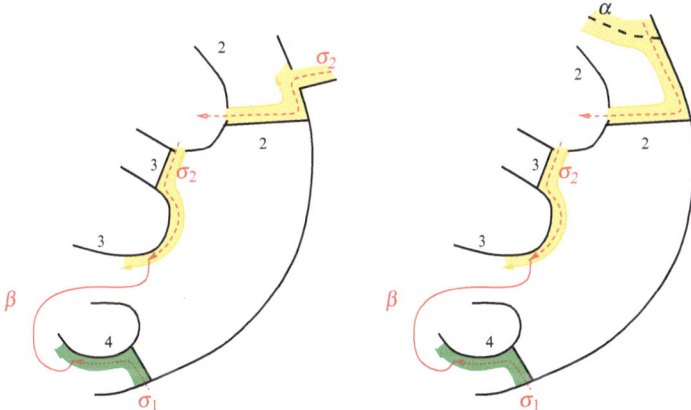

Fig. 4.5 The region R_W in the case that σ_2 does not immediately connect to β_W. *Left*: Initially σ_2 enters R_W by running upstream. *Right*: Initially σ_2 enters R_W along a non-prime arc. In both cases, σ_2 runs across a state circle C_2 and eventually re-enters R_W

Proof. If after entering R_W, σ_2 does not immediately meet β_W, then it must cross a state circle or non-prime arc first. It does not cross a non-prime arc, for if so, it would enter a non-prime half-disk bounded by the non-prime arc and some state circle C, so must exit this non-prime half-disk along C, and by Lemma 3.7 (Shortcut lemma), must do so running downstream. Then σ_2 must cross C again, to re-enter R_W, but then Lemma 3.11 (Utility lemma) implies it must first cross running upstream. This is impossible. So σ_2 does not cross a non-prime arc on the boundary of R_W between entering R_W and connecting to β_W.

Similarly, σ_2 cannot follow a tentacle downstream after crossing into R_W, or as above it would not be able to re-enter R_W. The only other possibility is that σ_2 follows a tentacle upstream, crossing into a state circle C_2. Then σ_2 must exit out of C_2. Lemma 3.11 (Utility lemma) implies that σ_2 exits C_2 in the downstream direction.

Now σ_2 is running downstream, so will be on the tail of a tentacle adjacent to a state circle C_3, attached to C_2 by a segment of H_A. Since σ_2 is running downstream, Lemma 3.10 (Downstream) implies it either continues running downstream, crossing into C_3, or crosses a non-prime arc α with W on the opposite side, or exits the shaded face to β_W. The first possibility cannot hold: σ_2 cannot cross C_3 running downstream, since it must cross out again to meet β_W, and this contradicts Lemma 3.11 (Utility lemma).

The second possibility will also lead to a contradiction. If σ_2 crosses a non-prime arc α with W on the opposite side, then α would have both endpoints on the state circle C_3. The non-prime half-disk containing W bounded by α and C_3 is therefore separated from the region containing C_W. But this is impossible: C_W meets the boundary of R_W. So the second possibility cannot hold either.

Fig. 4.6 Regions R_W in the case where σ_2 connects immediately to β_W

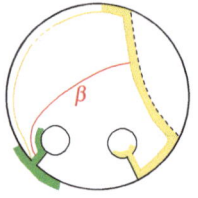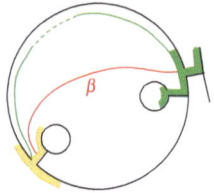

The only remaining possibility is that σ_2 immediately connects to β_W, as desired.

□

We are now ready to give the proof of Proposition 4.13: a normal square S whose white faces are in different polyhedral regions, which is glued to a normal square in a lower polyhedron, must parabolically compress.

Proof (of Proposition 4.13). As usual, let σ_1 and σ_2 be the arcs of the square S on the shaded faces. By Lemma 4.15 (Entering polyhedral region), we may assume that σ_1 enters R_W, the polyhedral region containing W, by crossing a state circle C_W in the downstream direction, and then immediately connects to β_W. We may also assume that σ_2 enters either in the upstream direction or across a non-prime arc.

Case 1: Suppose first that σ_2 also meets β_W immediately, without meeting any other boundary components of R_W. Then the region R_W and the arc β_W will have the form of one of the two graphs shown in Fig. 4.6, corresponding to the two possibilities for σ_2.

If the region marked X in the diagrams of Fig. 4.6 contains state circles, then if we push the endpoints of the arc β_W to the state circle on the outside in each diagram, we form a non-prime arc α, bounding state circles on either side. This contradicts the maximality of our prime decomposition, Definition 2.19. Thus the diagrams in Fig. 4.6 must contain no state circles in the regions marked X. Then in both cases, the tentacle running through X around the interior of the outermost state circle will terminate at the top of the tentacle where β_W has its other endpoint, as illustrated. Thus β_W cuts off a single ideal vertex of the white face W. But then the portion of the white disk bounded by these two shaded faces and the arc β_W forms a parabolic compression disk for S, as desired.

Case 2: Now suppose that σ_2 does not immediately connect to W after crossing C_W. Then Lemma 4.16 implies that the region R_W is as shown in Fig. 4.5. Recall that by assumption, S is glued to a square T in the lower polyhedron at W. Apply the clockwise map to β_W, sending it to an arc which differs from the arc of T lying in W by a clockwise rotation. The image of β_W is shown in Fig. 4.7.

Notice in the lower polyhedron that there exists an arc through two innermost disks (boundary components of R_W) adjacent to a single segment of H_A which connects the endpoints of the image of β_W under the clockwise map; see Fig. 4.7. In fact, this gives a normal trapezoid S' contained in the lower polyhedron with one of its sides on W, two sides on the two shaded faces corresponding to the

Fig. 4.7 Image of β_W under clockwise map $\phi\colon W \to W$. Note there is an arc (*dotted red*) running through two state circles and a single segment of H_A connecting its endpoints. All *red* arcs shown form a normal trapezoid in the lower polyhedron

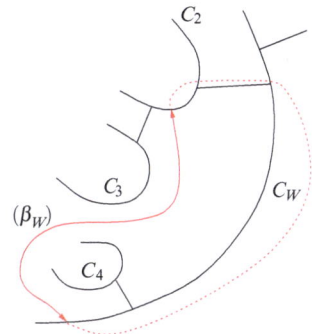

two innermost disks, and a side running over the ideal vertex of the polyhedron corresponding to the center of this edge of H_A.

Recall that we have the normal square T in the lower polyhedron with one side on the white face W, differing from the side of S' on W by a single clockwise rotation. Lemma 4.12 implies that S' and T are parabolically compressible to an ideal vertex of W. Thus, S is parabolically compressible at W. □

4.4 *I*-Bundles are Spanned by Essential Product Disks

We can now complete the proof of Theorem 4.4. Recall from the beginning of the chapter that the proof of Theorem 4.4 required three steps, whose proofs we have postponed until now. The first step, Proposition 4.18, relies on the following general lemma, which will also be needed in Chap. 5.

Lemma 4.17 (Product rectangle in white face). *Let B be an I-bundle in M_A, whose vertical boundary is essential. Suppose that B has been moved by isotopy to minimize its intersections with the white faces. Then, for any white face W, the intersection $B \cap W$ is a union of product rectangles whose product structure comes from the I-bundle structure of B. In other words, each component of $B \cap W$ has the form*

$$D = \alpha \times I,$$

where $\alpha \times \{0\}$ and $\alpha \times \{1\}$ are sub-arcs of ideal edges of W.

Proof. Suppose, first, that $B = Q \times I$ is a product I-bundle over an orientable base Q. At the end of the proof, we will consider the case of a non-orientable base.

Let D be a component of $B \cap W$. Notice that ∂D cannot contain any simple closed curves in the interior of W, because an innermost such curve would bound a compression disk for ∂B, and can be removed by isotopy. Similarly, ∂D cannot contain an arc from an ideal edge of W to the same ideal edge, since an outermost such arc can be removed by isotopy.

Truncate the ideal vertices of W, so that every ideal vertex becomes an arc (parallel to the parabolic locus). Abusing notation slightly, the portion of D in this truncated face is still denoted D. Then, by the above paragraph, D must be a $2n$-gon, with sides $\alpha_1, \beta_1, \ldots, \alpha_n, \beta_n$. Here, every α_i is a sub-arc of an ideal edge of W (and comes from the horizontal boundary of B), while every β_i is a normal arc that connects distinct ideal edges of W (and comes from the vertical boundary of B).

We claim that every β_i spans the product bundle $B = Q \times I$ top to bottom. For, suppose for concreteness that both endpoints of β_1 are on $Q \times \{1\}$. Then $\beta_1 \subset W$ is parallel to $Q \times \{1\}$ through $Q \times I$, which gives a boundary compression disk for the white face W. This contradicts Lemma 3.20 on p. 48, proving the claim. Note that this implies n is even.

Next, we claim that $n = 2$. For, suppose for a contradiction that $n > 2$. Then the sides α_1 and α_3 belong to the same (top or bottom) boundary of B, say $Q \times \{1\}$. There is an arc γ through the polygon $D \subset W$ that connects α_1 to α_3. This arc is parallel to $Q \times \{1\}$ through $Q \times I$, which again gives a boundary compression disk, contradicting Lemma 3.20. We conclude that D is a rectangle, with α_1, α_2 horizontal and β_1, β_2 vertical. Thus, after an appropriate isotopy, D is a vertical rectangle in the product structure on B.

Now, suppose that $B = Q \widetilde{\times} I$, where Q is non-orientable. Let $\gamma_1, \ldots, \gamma_n$ be a maximal collection of disjoint, embedded, orientation-reversing loops in Q. Then the I-bundle over each γ_i is a Möbius band A_i. Furthermore, $Q \setminus (\cup \gamma_i)$ is an orientable surface Q_0, such that $B_0 = B \setminus (\cup A_i)$ is a product I-bundle over Q_0. Let W be a white face, and let D be a component of $B \cap W$. By the orientable case already considered, every component of $B_0 \cap W$ is a product rectangle $\alpha \times I$. Also, $A_i \cap W$ is a union of arcs, hence the regular neighborhood of each vertical Möbius band A_i intersects W in rectangular product strips. Each of these strips respects the I-bundle structure of B. Thus D is constructed by joining together several product rectangles of $B_0 \cap W$, along product rectangles in the neighborhood of $A_i \cap W$. Therefore, all of D is a product rectangle, as desired. $\square$

Proposition 4.18 (Step 1). *Let B be a non-trivial I-bundle of the characteristic submanifold of M_A. Then B meets the parabolic locus of M_A.*

Proof. Recall that since B is non-trivial we have $\chi(B) < 0$. By Lemma 4.6, B contains a product bundle $Y = Q \times I$, where Q is a pair of pants. Moreover, when put into normal form in a prime decomposition of M_A, the three annuli of ∂Y are composed of disjointly embedded normal squares. Label the squares $S_1, \ldots, S_n$. Observe each has the form of a product $S_i = \gamma_i \times I$, where γ_i is a sub-arc of ∂Q.

If some S_i is parabolically compressible, observe that the parabolic compression disk E that connects S_i to the parabolic locus is itself a product I-bundle, which can be homotoped to have product structure matching that of S_i. Hence $E \subset B$, and B borders the parabolic locus, as desired.

If Y passes through more than one lower polyhedron, some square S_i must pass through white faces in different polyhedral regions. Thus, by Proposition 4.13, S_i is parabolically compressible. Hence, as above, B borders the parabolic locus.

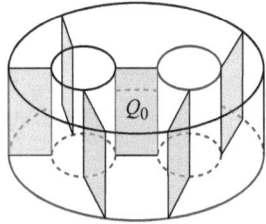 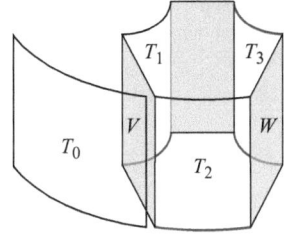

Fig. 4.8 *Left*: the product bundle $Q \times I$ for a pair of pants Q. *Right*: the prism $Q_0 \times I$

For the rest of the proof, assume that every S_i is parabolically incompressible, and so Y is entirely contained in the upper polyhedron and exactly one lower polyhedron P. This assumption will lead to a contradiction.

Consider the intersections between $Y = Q \times I$ and the white faces. By Lemma 4.17, each component of intersection is a product rectangle $\alpha \times I$, where α is an arc through the interior of Q. Thus Y intersects the individual polyhedra in a finite number of prisms, each of which is the product of a polygon with an interval. Vertical faces of the prism alternate between product rectangles on white faces and normal squares S_i. Let $Y_0 = Q_0 \times I$ be the prism whose base polygon has the greatest number of sides. Since Q is a pair of pants, and has negative Euler characteristic, Q_0 must have at least six sides, half of which lie on normal squares S_i. See Fig. 4.8.

Let $T_1, \ldots, T_k$ denote the normal squares that bound Y_0, listed in order. By the above, $k \geq 3$. Let V be the white face containing the rectangle of Y_0 between T_1 and T_2, and let W be the white face containing the rectangle between T_2 and T_3. Finally, let T_0 denote the normal square of ∂Y glued to T_2 at the white face V. Thus each T_j is one of the S_i, relabeled. Note that if Y_0 is contained in the upper polyhedron, then so are $T_1, \ldots, T_k$, but T_0 is contained in the lower polyhedron. Similarly, if Y_0 is in the lower polyhedron, then so are $T_1, \ldots, T_k$, but T_0 is in the upper polyhedron.

Using Lemma 4.8, map all the T_j to the lower polyhedron, by the clockwise map. If T_0 is in the upper polyhedron, let T_0' be its image under the clockwise map. Otherwise, if T_0 is in the lower polyhedron, let $T_0' = T_0$. Similarly, if T_i, $1 \leq i \leq k$ is in the upper polyhedron, let T_i' be its image under the clockwise map. Otherwise, let $T_i' = T_i$. Notice that because $T_1, \ldots, T_k$ are pairwise disjoint, Lemma 4.8(3) implies that $T_1', \ldots, T_k'$ must also be disjointly embedded in the lower polyhedron P.

Now, since T_0 is glued to T_2 across the white face V, T_0' must differ from T_2' by a single rotation of V. Since we are assuming that T_0' and T_2' cannot be parabolically compressible at V, Lemma 4.10 implies T_0' and T_2' must intersect, both in V and in the other white face met by T_2', which recall is the face W.

Now, T_1' runs parallel to T_2' through V. Thus T_0' must also intersect T_1', and again Lemma 4.10 implies that T_0' and T_1' intersect in W. Similarly, T_3' runs parallel to T_2' through W, so Lemma 4.10 implies that T_0' intersects T_3' in W and in V. But now, T_1', T_2', and T_3' are disjoint normal squares through V and W, so must be parallel, and in particular, T_2' must separate T_1' and T_3'. See Fig. 4.9. On the other hand, T_1',

Fig. 4.9 Normal squares in
the lower polyhedron that
occur in the proof of
Proposition 4.18

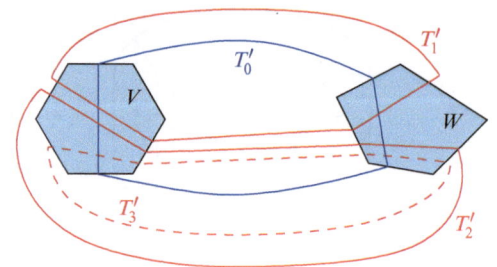

T_2' and T_3' are all lateral faces of the same contractible prismatic block Y_0. This is a
contradiction.   □

The following proposition supplies Step 2 of the proof. We note that the proof of
the proposition is a straightforward topological argument that doesn't use any of the
machinery we have built.

Proposition 4.19 (Step 2). *Suppose B is a non-trivial connected component of the
maximal I-bundle of M_A, which meets the parabolic locus. Then B is spanned by
essential product disks.*

Proof. Since B is a 3-dimensional submanifold of S^3, it is orientable. Hence B is
either the product I-bundle over an orientable surface F, or the twisted I-bundle
over a non-orientable surface F. In either case, $\chi(F) < 0$. Since B meets the
parabolic locus of M_A, so does F.

We may fill F with disjoint edges with endpoints on the parabolic locus of F,
subdividing F into disjoint triangles. Now consider the set of points lying over any
such edge of the triangulation. This will be an essential product disk, meeting the
parabolic locus of M_A at those points that lie over the endpoints of the edge, and
meeting S_A elsewhere. Remove all such essential product disks from B.

In F, removing such arcs gives a finite collection of open disks. The I-bundle
over such a disk is a prism over a triangle, so we have satisfied the definition of
spanning, Definition 4.3.   □

To complete the third and final step of the proof of Theorem 4.4, we need the
following lemma.

Lemma 4.20. *Suppose S is a normal disk in the upper polyhedron, glued to a
normal disk T in the lower polyhedron along a white face W. If one of S or T is
a normal trapezoid and the other is either a normal trapezoid or a normal square,
then both S and T parabolically compress at W.*

Proof. For each of S and T, define normal squares S' and T' in the following way.
If S is a normal square, then let $S' = S$. If S is a trapezoid, then let S' be the
normal square obtained by pulling S off the parabolic locus, and into a white face
V. Similarly, if T is a normal square, we let $T' = T$; otherwise, if T is a trapezoid,
we define T' to be the normal square obtained by pulling T off the parabolic locus.

Notice that the resulting squares S' and T' are glued to each other at W. By Proposition 4.13, if V and W do not belong to the same region of $s_A \cup (\cup_i \alpha_i)$, S' is parabolically compressible at W. If S' parabolically compresses at W, then so does T' (because it is glued to S' at W).

Thus, we may assume that V and W belong to the same region of the complement of $s_A \cup (\cup_i \alpha_i)$. Then the entire boundary of S' can be mapped to the boundary of a square S'' in the lower polyhedron containing T', via the clockwise map of Definition 4.7 and Lemma 4.8.

By hypothesis, either S or T (or both) is a trapezoid. Thus either S'' or T' cuts off a single ideal vertex in a white face other than W, hence S'' and T do not intersect in any white face other than W. This means we must have conclusion (1) of Lemma 4.10: S'' and T' do not intersect at all, and each of them cuts off an ideal vertex of W. Therefore, both S and T are parabolically compressible at W.   □

Proposition 4.21 (Step 3). *Let D be an essential product disk in M_A. Then D parabolically compresses to a collection of essential product disks, each of which is embedded in a single polyhedron.*

Proof. Let D be an essential product disk in M_A. If D is disjoint from all white faces, then we are done: D is contained in a single polyhedron.

If D meets white faces, then they split it into disks $S_1, \ldots, S_n$. Because our polyhedra cannot contain any normal bigons (Lemma 3.18), every arc of intersection between D and a white face must run from one side of D to the opposite side. Thus S_1 and S_n are normal trapezoids, and $S_2, \ldots, S_{n-1}$ are normal squares. Consider S_1 and S_2. One of these is in the upper polyhedron, and one in the lower. They meet at a white face W. Lemma 4.20 implies that both S_1 and S_2 must parabolically compress at W. So D compresses to essential product disks D_1 and D_2, where D_1 is the compressed image of S_1 and D_2 is the compressed image of $S_2 \cup \cdots \cup S_n$. Repeat this argument for the essential product disk D_2. Continuing in this manner, we see that D parabolically compresses to essential product disks, each in a single polyhedron.   □

4.5   The σ-Adequate, σ-Homogeneous Setting

All the results of this chapter also hold in the setting of the ideal polyhedral decomposition for σ-adequate, σ-homogeneous diagrams. Here, we briefly discuss this generalized setting.

Lemma 4.6 holds, and the proof requires no changes. Hence in the characteristic I-bundle, we may find a product bundle $Y = Q \times I$, where Q is a pair of pants and the annuli of ∂Y are composed of embedded normal squares.

In Definition 4.7, the *clockwise map* was defined to map faces of the upper polyhedron to faces of the lower. Each of our white faces in the σ-homogeneous case is contained in an all-B or all-A polyhedral region. In the latter case, we use the same clockwise map as before. As the map is identical, all the results in that section hold. In the all-B case, rather than mapping by one rotation in the clockwise direction, we need to map by one rotation in the counter-clockwise direction. However, the properties in Lemma 4.8 will still hold in the all-B case, and the proof goes through without change.

We then have Lemma 4.9, which discusses the intersections of normal squares in a checkerboard polyhedron. This lemma is due to Lackenby [58, Lemma 7], and holds in complete generality. This immediately implies Lemma 4.10: two normal squares with arcs in the same white face which differ by a single rotation, will either each cut off a single ideal vertex in that face and not intersect at all, or intersect nontrivially in both of their corresponding white faces. We also obtain Lemma 4.12.

The results of Sect. 4.3 will hold as well. A check through their proofs indicates that they require the named lemmas from Chap. 3, which we have shown to hold in the σ-adequate, σ-homogeneous case. In particular, Proposition 4.13 holds: A normal square whose white faces are in different polyhedral regions, glued to a normal square in a lower polyhedron, must parabolically compress. More particularly, it will cut off a single ideal vertex in the white face. The proof of the proposition uses Lemmas 4.15 and 4.16, as well as Lemma 4.12, which continue to hold. There are two cases of the proof. The second uses the clockwise map. In the case that the polyhedral region is all-B, we must use the "counter-clockwise map" instead. This requires reflecting the figures that illustrate the proof, but the combinatorics of the situation will remain unchanged.

Finally, we step through the results of Sect. 4.4. Lemma 4.17 (Product rectangle in white face) requires only Lemma 3.20, which follows immediately from Proposition 3.24 (No normal bigons) in the σ-adequate, σ-homogeneous case. Hence it continues to hold in this setting. Similarly, Proposition 4.18 holds. The proof applies verbatim, with the sole modification that the "clockwise map" must be replaced by the "counter-clockwise map" in an all-B polyhedral region.

Proposition 4.19 holds without change. Lemma 4.20 holds after replacing "clockwise" with "clockwise or counter-clockwise" in the proof. Finally, Proposition 4.21 holds without change.

Thus all the results of this chapter hold for σ-adequate, σ-homogeneous diagrams. In particular, the following generalization of Theorem 4.4 reduces the problem of understanding the I-bundle of the characteristic submanifold of $M_\sigma = S^3 \backslash\backslash S_\sigma$ to the problem of understanding and counting EPDs in individual polyhedra.

Theorem 4.22. *Let D be a (connected) diagram of a link K, and let σ be an adequate, homogeneous state of D. Let B be a non-trivial component of the characteristic submanifold of $M_\sigma = S^3 \backslash\backslash S_\sigma$. Then B is spanned by a collection of essential product disks $D_1, \ldots, D_n$, with the property that each D_i is embedded in a single polyhedron in the polyhedral decomposition of M_σ.*   □

Chapter 5
Guts and Fibers

This chapter contains one of the main results of the manuscript, namely a calculation of the Euler characteristic of the guts of M_A in Theorem 5.14. The calculation will be in terms of the number of essential product disks (EPDs) for M_A which are *complex*, as in Definition 5.2, below. In subsequent chapters, we will find bounds on the number of such EPDs in terms of a diagram, for general and particular types of diagrams (Chaps. 6–8), and use this information to bound volumes, and relate other topological information to coefficients of the colored Jones polynomial (Chap. 9).

Recall that we have shown in Theorem 4.4 that the I-bundle of M_A is spanned by EPDs, each of which is embedded in a single polyhedron of the polyhedral decomposition. (See Definitions 4.2 and 4.3 on p. 55 to recall the terminology.) Thus to calculate the Euler characteristic of the guts, we calculate the minimal number of such a collection of spanning EPDs. We will do this by explicitly constructing a spanning set of EPDs with desirable properties (Lemmas 5.6 and 5.8). In Proposition 5.13, we will compute exactly how redundant the spanning set is. This leads to the Euler characteristic computation in Theorem 5.14. Along the way, we also give a characterization of when the link complement fibers over S^1 with fiber the state surface S_A, in terms of the reduced state graph $\mathbb{G}'_A$, in Theorem 5.11.

5.1 Simple and Non-simple Disks

By Theorem 4.4, every non-trivial component in the characteristic submanifold of M_A is spanned by essential product disks in individual polyhedra. Our goal is to find and count these disks, starting with the lower polyhedra.

Lemma 5.1. *Let D be an A-adequate diagram of a link in S^3. Consider a prime polyhedral decomposition of $M_A = S^3 \backslash\backslash S_A$. The essential product disks embedded in the lower polyhedra are in one-to-one correspondence with the 2-edge loops in the graph $\mathbb{G}_A$.*

D. Futer et al., *Guts of Surfaces and the Colored Jones Polynomial*, Lecture Notes in Mathematics 2069, DOI 10.1007/978-3-642-33302-6_5,
© Springer-Verlag Berlin Heidelberg 2013

Proof. By definition, an EPD in a lower polyhedron must run over a pair of shaded faces F and F'. By Lemma 2.21 on p. 29, these shaded faces correspond to state circles C and C'. Furthermore, every ideal vertex shared by F and F' corresponds to a segment of H_A between C and C', or equivalently, to an edge of $\mathbb{G}_A$ between C and C'. Since an EPD must run over two ideal vertices between F and F', it naturally defines a 2-edge loop in $\mathbb{G}_A$, whose vertices are the state circles C and C'. In the other direction, the two edges of a 2-edge loop in $\mathbb{G}_A$ define a pair of ideal vertices shared by F and F', hence an EPD. Thus we have a bijection. $\square$

Typically, we do not need *all* the disks in the lower polyhedra to span the I-bundle. We will focus on choosing disks that are as simple as possible.

Definition 5.2. Let P be a checkerboard-colored ideal polyhedron. An essential product disk $D \subset P$ is called

(1) *Simple* if D is the boundary of a regular neighborhood of a white bigon face of P,
(2) *Semi-simple* if D parabolically compresses to a union of simple disks (but is not itself simple),
(3) *Complex* if D is neither simple nor semi-simple.

For example, in Fig. 4.2 on p. 56, the disk on the left is semi-simple, and the disks on the right are simple.

In certain special situations (for example, alternating diagrams studied by Lackenby [58] and Montesinos diagrams studied in Chap. 8), simple disks suffice to span the I-bundle of M_A. In general, however, we may need to use complex disks.

Example 5.3. Consider the A-adequate link diagram shown in Fig. 5.1, left. The graph H_A for the diagram is shown in the center of the figure. Note that there are exactly 3 polyhedral regions, hence exactly 3 lower polyhedra. In each polyhedral region, there is exactly one 2-edge loop of $\mathbb{G}_A$. Thus, by Lemma 5.1, there are exactly 3 EPDs in the lower polyhedra. Two of these (in the innermost and outermost polyhedral regions) are simple by Definition 5.2, and may be isotoped through bigon faces into the upper polyhedron. However, the green and orange[1] shaded faces of the upper polyhedron shown in the right panel of Fig. 5.1 meet in a total of six ideal vertices. Since an EPD connects two ideal vertices, and a set of EPDs that spans the part of the I–bundle contained in the upper polyhedron must connect all six vertices, a minimum of five EPDs are required to span the part of the I-bundle contained in the upper polyhedron. This requires using complex EPDs, for example the ones shown in Fig. 5.1.

One feature of this example is that modifying the link diagram fixes the problem. The modified link diagram in Fig. 5.2 is still A-adequate. This time, all the EPDs in the lower polyhedra are simple or semi-simple. Furthermore, simple EPDs (isotoped across bigon faces from the lower polyhedra into the upper) account for all the ideal

---

[1]Note: For grayscale versions of this monograph, green will refer to the darker gray shaded face, orange to the lighter one.

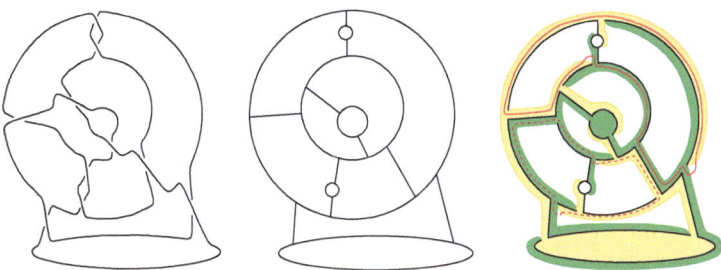

Fig. 5.1 *Left*: an *A*-adequate link diagram. *Center*: its graph H_A. *Right*: shaded faces in the upper polyhedron, with (normal squares corresponding to) two if an essential disk *D* complex EPDs shown in *red*

Fig. 5.2 *Left*: an alternate
A-adequate diagram of the
link in Fig. 5.1. *Right*: the
graph H_A for this diagram

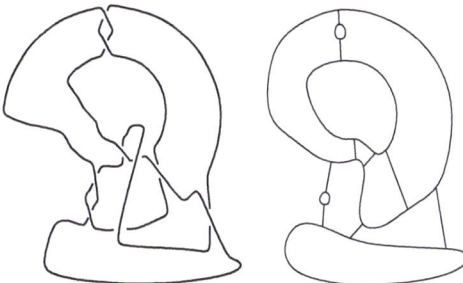

vertices in the upper polyhedron where an EPD may cross from one shaded face into another. Thus, in the modified diagram, simple EPDs suffice to span the *I*-bundle. This phenomenon of modifying a diagram to remove complex EPDs is discussed again in Chap. 10.

Definition 5.4. Let *D* be an essential product disk in a polyhedron *P*. Since *P* is a ball, *D* separates *P* into two sides. We say that *D* is *parabolically incompressible to one side* (or *PITOS* for short) if all parabolic compression disks for *D* lie on the same side of *D*.

Note that simple disks, which have a bigon face to one side, are automatically PITOS.

A convenient way to characterize PITOS disks is via the following lemma.

Lemma 5.5. *Let P be a checkerboard-colored ideal polyhedron, and let F and G be shaded faces of P. Let $v_1, \ldots, v_n$ be the ideal vertices at which F meets G. Then*

(1) *If we label $v_1, \ldots, v_n$ such that the vertices are ordered consecutively around the boundary of F, for example according to a clockwise orientation on $\partial F = S^1$, then $v_1, \ldots, v_n$ will also be ordered consecutively on ∂G, but with the reverse orientation (counterclockwise).*
(2) *With the consecutive ordering of (1), an essential product disk, running through faces F and G and ideal vertices v_i and v_j, is PITOS if and only if $j = i \pm 1$ (mod n).*

(3) *If $n \geq 3$ and every PITOS disk through faces F and G is simple, then F and G are the only shaded faces of P.*

Proof. We may identify ∂P with $S^2 \cong \mathbb{R}^2 \cup \{\infty\}$, in such a way that ∞ falls in the interior of a white face. Then the orientation on $\mathbb{R}^2$ induces a (clockwise) orientation on the boundary of every shaded face of P.

Let n be the number of ideal vertices at which F meets G. If $n < 2$, then there are no EPDs through the pair of faces F, G, and the claims of the lemma are trivial. Thus we may assume $n \geq 2$.

Order these vertices $v_1, \ldots, v_n$, clockwise around the boundary of F. For conclusion (1), we claim that the ideal vertices $v_1, \ldots, v_n$ are ordered counterclockwise around the boundary of G.

Let v_i and v_{i+1} be vertices that are consecutive on ∂F. Then there is an essential product disk that runs through F and G, and meets exactly these ideal vertices. Let γ be the boundary of this disk. By the Jordan curve theorem, γ cuts $\mathbb{R}^2$ into an inside and an outside region. Let α_F be the oriented (clockwise) arc of ∂F from v_i to v_{i+1}. Without loss of generality, α_F lies inside γ.

Now, consider the portion of G that lies inside γ. If this portion of G has any ideal vertices meeting F, they would have to meet F inside γ. But, by construction, the portion of ∂F inside γ is a single arc α_F, without any additional ideal vertices. Therefore, inside γ, G has no vertices meeting F. Hence, v_i and v_{i+1} must be consecutive from the point of view of G. The orientation on the plane means that the clockwise arc $\alpha_G \subset \partial G$ that lies inside γ must run from v_{i+1} to v_i. Thus the vertices $v_1, \ldots, v_n$ are in counterclockwise order around ∂G, proving (1).

For (2), observe that if an essential disk D runs through consecutive vertices v_i and v_{i+1}, then all other vertices shared by F and G are on the same side. Thus all parabolic compressions of D are on the same side, and D is PITOS. Conversely, if v_i and v_j are not consecutive, then there are parabolic compressions on both sides, and D is not PITOS.

It remains to show (3). By Definition 5.2, any simple disk D through F and G is parallel to a white bigon face of P. When $n \geq 3$, one component of $P \setminus \partial D$ contains an extra ideal vertex, hence cannot be a bigon. The bigon face must be on the other side, which we call the *inside* of D. Thus, when $n \geq 3$ and all PITOS disks through F and G are simple, the insides of these disks are disjoint.

Under these hypotheses, we have mapped out the entire polyhedron P. Inside each of the n essential product disks is a white bigon face, with no extra vertices. Each essential product disk meets F in an arc and G in an arc. Thus outside all these disks, there is an n-gon in F containing no additional ideal vertices, meeting an n-gon in G containing no additional ideal vertices, where the n-gons meet at their vertices. Since there are no additional vertices, there can be no additional faces, white or shaded. Thus F and G are the only shaded faces in the polyhedron P. $\square$

5.2   Choosing a Spanning Set

Next, we will construct a spanning set for the part of the I-bundle that is contained in each individual polyhedron. By Definition 4.3, a collection of EPDs span the I-bundle of M_A if the complement of these disks in the I-bundle is a union of prisms and solid tori. Because our goal is to count the Euler characteristic of the I-bundle, prisms and solid tori are counted differently. As we construct the spanning set, we will keep track of the number of prisms created.

Recall that a lower polyhedron P corresponds to a polyhedral region of the diagram. Let $e_A(P)$ be the number of segments of H_A (equivalently, the number of edges of $\mathbb{G}_A$) in this polyhedral region, and $e'_A(P)$ be the number of reduced edges (after duplicates are removed). We may now choose a spanning set of EPDs for the polyhedron P.

Lemma 5.6. *Let P be a lower polyhedron in the polyhedral decomposition of M_A. Then all the essential product disks in P are spanned by a particular spanning set $E_l(P)$, with the following properties:*

(1) *Every simple disk in P belongs to the spanning set $E_l(P)$.*
(2) *No disks in $E_l(P)$ are semi-simple. (Recall semi-simple disks are not simple by definition.)*
(3) *The cardinality of $E_l(P)$ is $\|E_l(P)\| = e_A(P) - e'_A(P) + \varepsilon_P$, where ε_P is either 0 or 1.*
(4) *The following are equivalent:*

    (a) P has exactly two shaded faces.
    (b) All white faces of P are bigons.
    (c) $P\backslash\backslash E_l(P)$ contains a prism over an ideal n-gon.
    (d) $e'_A(P) = 1$, and this single edge separates the graph $\mathbb{G}'_A$.
    (e) $\varepsilon_P = 1$.

Proof. We construct the spanning set as follows. For every pair of shaded faces F, G of the polyhedron P, let $E_{F,G}$ be the set of all PITOS essential product disks that run through F and G. If the number of ideal vertices shared by F and G is n, then $E_{F,G}$ will be non-empty precisely when $n \geq 2$. By Lemma 5.5, these disks are in 1–1 correspondence with consecutive pairs of vertices v_i, v_{i+1} shared by F and G.

Now, we consider two cases.

Case 1: F and G are the only shaded faces in P. In this case, we let $E_l(P) = E_{F,G}$.

Let us check the conclusions of the lemma. Note that every simple disk is PITOS, hence must belong to $E_l(P) = E_{F,G}$. Conversely, every disk in $E_{F,G}$ is PITOS, hence contains no vertices between F and G on one side, hence contains no vertices at all on that side, and thus can only be simple.

Recall that the shaded faces F and G correspond to state circles C_F and C_G. Since these are the only shaded faces of P, then C_F and C_G are the only state circles in the polyhedral region of P. All the edges of $\mathbb{G}_A$ in the polyhedral region must connect C_F and C_G; there are $n = e_A(P)$ such edges total. In the reduced graph $\mathbb{G}'_A$, these n edges are identified to one, hence $e'_A(P) = 1$. Thus

$$\|E_l(P)\| = n = e_A(P) - e'_A(P) + \varepsilon_P, \quad \text{where} \quad \varepsilon_P = 1.$$

In Case 1, all the conditions of (4) will be true. The polyhedron P has exactly two shaded faces F and G, and all the white faces are bigons parallel to the simple disks. Cutting P along the disks of $E_l(P)$ produces a prism over an ideal n-gon. We have already seen that $\varepsilon_P = e'_A(P) = 1$. Finally, because every state circle in S^2 is separating, any path in H_A between state circles C_F and C_G must pass through the polyhedral region of P, hence must use one of the n edges that are identified to one in $\mathbb{G}'_A$.

Case 2: F and G are not the only shaded faces in P.

When F and G share n vertices with $n \geq 2$, we will see that we may remove one of the n disks in $E_{F,G}$ to obtain a set $E'_{F,G}$ of $(n-1)$ disks, which still span all the EPDs through faces F and G. We make the choices as follows. If $n = 2$, then the two disks $E_{F,G}$ both run through vertices v_1 and v_2, and are parallel. So we may omit one. If $n \geq 3$, then Lemma 5.5 implies that one of the disks in $E_{F,G}$ is non-simple. Thus we omit a non-simple disk. Note that by construction, one copy of each simple disk through faces F and G remains in $E'_{F,G}$. Note further that there is a prism between all disks of $E'_{F,G}$, so the removed disk is spanned by the remaining ones. Since all EPDs through faces F and G are spanned by PITOS ones, the remaining set of $(n-1)$ disks spans all EPDs through F and G, as claimed.

When F and G share fewer than 2 ideal vertices, $E_{F,G}$ is empty, and for notational convenience we set $E'_{F,G}$ to be empty.

Let us check that in the non-trivial cases, all disks in $E'_{F,G}$ satisfy conclusion (2): that is, none of them is semi-simple. If $n = 2$, there is nothing to check, because no parabolic compressions are possible. Thus, suppose that $n \geq 3$, and we obtained $E'_{F,G}$ by omitting a non-simple disk in $E_{F,G}$.

Suppose for a contradiction that $D \in E'_{F,G}$ is not simple, but parabolically compresses to simple disks. Because the ideal vertices v_i, v_{i+1} met by D are consecutive, the parabolic compression must be to the *outside* of D—that is, away from the arc of ∂F that runs from v_i to v_{i+1}. Any simple disks to which D compresses must be PITOS, hence belong to $E_{F,G}$. But one of the disks to the outside of D is not simple, contradicting the hypothesis that D compresses to simple disks to its outside.

Now, let $E_l(P)$ be the union of all the sets $E'_{F,G}$, as (F, G) ranges over all unordered pairs of shaded faces in P. We have already checked that this spanning set satisfies conclusions (1) and (2).

Observe that the shaded faces F and G correspond to state circles C_F and C_G. If there are two or more edges of $\mathbb{G}_A$ connecting C_F to C_G (i.e., if $n \geq 2$), then some of these edges will be removed as we pass to the reduced graph $\mathbb{G}'_A$. The number of edges removed is exactly $n - 1$, which is equal to the cardinality of $E'_{F,G}$. Thus

$$||E_l(P)|| = \sum_{(F,G)} ||E'_{F,G}|| = e_A(P) - e'_A(P) + \varepsilon_P, \quad \text{where} \quad \varepsilon_P = 0.$$

The sum is over unordered pairs of shaded faces (F, G).

In Case 2, all the conditions of (4) will be false. By hypothesis, the polyhedron P has more than two shaded faces, hence some white face is not a bigon. Since the polyhedral region of P has more than two state circles and is connected, this region must contain more than one edge of G'_A. The one non-trivial statement in (4) is that $P\backslash\backslash E_l(P)$ cannot contain a prism.

Suppose, for a contradiction, that $P\backslash\backslash E_l(P)$ contains a prism over an n-gon. Then the top and bottom faces of this prism are on shaded faces F and G of P, and the lateral faces are EPDs in $E_l(P)$. By construction, these lateral faces must belong to $E'_{F,G}$. But then one of these lateral EPDs must parabolically compress to the remaining $(n - 1)$ EPDs—which is impossible, since we removed the one redundancy in $E_{F,G}$ when constructing the set $E'_{F,G}$. □

Lemma 5.6 has the following immediate consequence.

Lemma 5.7. *Let E_l be the union of all the spanning sets $E_l(P)$, as P ranges over all the lower polyhedra. Then every essential product disk in one of the lower polyhedra is spanned by the disks in E_l. The set E_l contains all simple disks in the lower polyhedra. Furthermore,*

$$||E_l|| = e_A - e'_A + n_{\text{sep}},$$

where n_{sep} is the number of separating edges in G'_A, and n_{sep} is also equal to the number of prisms in the lower polyhedra in the complement of E_l.

Proof. The properties that E_l contains all simple disks and spans all the EPDs in the lower polyhedra follow immediately from the same properties of the constituent sets $E_l(P)$. To compute the cardinality of E_l, it suffices to observe that the total number of edges removed as we pass from G_A to G'_A is

$$e_A - e'_A = \sum_P e_A(P) - e'_A(P),$$

and the total contribution of the terms ε_P in Lemma 5.6 is exactly n_{sep}. By Lemma 5.6, these n_{sep} edges are in one-to-one correspondence with prisms in the lower polyhedra in the complement of E_l. □

Finally, we choose a spanning set of EPDs for the upper polyhedron.

Lemma 5.8. *Let P denote the upper polyhedron in the decomposition of M_A. Then there exists a set $E_s \cup E_c$ of essential products disks embedded in P, such that the following hold:*

(1) *E_s is the set of all simple disks in P.*
(2) *E_c consists of complex disks. Furthermore, E_c is minimal, in the sense that no disk in E_c parabolically compresses to a subcollection of $E_s \cup E_c$.*
(3) *The set $E_s \cup E_c$ spans the essential product disks in P.*
(4) *The following are equivalent:*

  (a) *$\mathbb{G}'_A$ is a tree.*
  (b) *Every white face is a bigon.*
  (c) *$P \backslash\backslash (E_s \cup E_c)$ contains exactly one prism.*
  (d) *Every (upper or lower) polyhedron is a prism, with horizontal faces shaded and lateral faces white.*

Proof. The construction is identical to the construction in Lemma 5.6. For every pair of shaded faces F and G, we let $E_{F,G}$ be the set of all PITOS disks that run through F and G. If F and G are the only shaded faces of the upper polyhedron P, we let $E_s = E_{F,G}$. In this case, all white faces of P are bigons, and all disks in $E_{F,G}$ are simple. Hence, $E_c = \emptyset$.

Alternately, if F and G are not the only shaded faces of P, we proceed as in Case 2 of Lemma 5.6. We prune the set $E_{F,G}$ by one disk, while keeping all simple disks, to obtain $E'_{F,G}$. As in the proof of Lemma 5.6, no disk in $E'_{F,G}$ is semi-simple. Then, we let $E_s \cup E_c$ be the union of all sets $E'_{F,G}$ as F, G range over the shaded faces of the polyhedron P. This combined set is composed of simple disks in E_s and complex disks in E_c.

In either case, we have constructed a set $E_s \cup E_c$ that satisfies conclusions (1) and (3).

To prove (2), observe that by construction, each disk in E_c is complex and PITOS. Suppose, for a contradiction, that some disk $D \in E_c$ parabolically compresses to other disks in $E_s \cup E_c$. Then, D would need to compress to the remaining $(n-1)$ PITOS disks that share the same shaded faces F and G (where n is the number of vertices at which F and G meet). But by construction, the only scenario in which all n disks of $E_{F,G}$ remain in $E_s \cup E_c$ is when all of these disks are simple, hence $E_c = \emptyset$, which is a contradiction.

It remains to prove the equivalent conditions of (4).

(4a) $\Leftrightarrow$ (4b): The connected graph $\mathbb{G}'_A$ is a tree if and only if every edge separates. Hence, this equivalence is immediate from Lemma 5.6(4).

(4b) $\Rightarrow$ (4d): Let P_0 be any polyhedron in the decomposition, and suppose that every white face of P_0 is a bigon. Then the white faces of P_0 must line up cyclically end to end, and there are exactly two shaded faces. Since a bigon white face is the product of an ideal edge with I, this product structure extends over the entire polyhedron. Thus P_0 is a prism whose horizontal faces are shaded and whose lateral faces are white bigons.

(4d) $\Rightarrow$ (4c): Suppose the top polyhedron P is a prism, whose lateral faces are white bigons. Parallel to every bigon face of P is a simple essential product disk, and by property (1), each of these simple EPDs is in the spanning set E_s. Thus $P\backslash\backslash E_s$ consists of a product region parallel to each white face, as well as a prism component separated from all white faces.

(4c) $\Rightarrow$ (4b): Suppose that $P\backslash\backslash (E_s \cup E_c)$ contains a prism over an n-gon. Then the top and bottom faces of this prism are on shaded faces F and G of P, and the lateral faces are EPDs in $E_{F,G}$. Notice that one of these lateral EPDs must parabolically compress to the remaining $(n-1)$ EPDs. This would be impossible if we removed one of the n disks in $E_{F,G}$ while passing to the reduced set $E'_{F,G}$. Thus every disk of $E_{F,G}$ must belong to $E_s \cup E_c$, which means that F and G are the only shaded faces in polyhedron P. Therefore, every white face of P is a bigon. But since every white face of a lower polyhedron is glued to P, all white faces must be bigons. $\square$

Definition 5.9. The spanning set E_c is defined in the statement of Lemma 5.8. Note that by Lemma 5.8(2), the cardinality $||E_c||$ is the smallest number of complex disks required to span the I-bundle of the upper polyhedron. We may take this property to be a definition of $||E_c||$.

Since the polyhedral decomposition is uniquely specified by the diagram $D(K)$ (see Chap. 2 and Remark 3.17), $||E_c||$ is a diagrammatic quantity, albeit one that is not easy to eyeball. In Chap. 7, we will bound the quantity $||E_c||$ in terms of simpler diagrammatic quantities, and in Chap. 8, we will prove that for most Montesinos links, $||E_c|| = 0$.

We also record the following property of the spanning set $E_s \cup E_c$, which will be needed in Chap. 7.

Lemma 5.10. *Let F and G be shaded faces of the upper polyhedron P, and let $E_s \cup E_c$ be the spanning set of Lemma 5.8. Then, for every tentacle of F, at most two disks of $E_s \cup E_c$ run through F and G and intersect this tentacle.*

Proof. Let α be an arc that cuts across a tentacle of F. Thus, from the point of view of P, α is an arc from an ideal vertex w to a point x in the interior of a side of P.

Let $v_1, \ldots, v_n$ be the ideal vertices shared by F and G, labeled in order, as in Lemma 5.5(1). The point $x \in \partial F \cap \alpha$ falls between a consecutive pair of vertices v_i that connect F to G. Then no generality is lost in assuming that x lies on the oriented arc from v_n to v_1.

Recall that all disks in $E_s \cup E_c$ are PITOS, and that by Lemma 5.5, PITOS disks through F and G must meet consecutive ideal vertices. The proof will be complete once we show that α can only meet at most two such disks (up to isotopy).

If w is not one of the vertices at which F meets G, then it lies between vertices v_i and v_{i+1}. In this case, α partitions $\{v_1, \ldots, v_n\}$ into two subsets: namely, $\{v_1, \ldots, v_i\}$ and $\{v_{i+1}, \ldots, v_n\}$. Any disk through F and G whose vertices belong to the same subset will be disjoint from α (up to isotopy). Thus α can only intersect the two disks that run from v_i to v_{i+1} and from v_n to v_1.

If w is one of the vertices v_i, then the argument is the same. In this case, α can only intersect the disk that runs from v_n to v_1.   □

5.3 Detecting Fibers

In this section, we prove that the Euler characteristic $\chi(\mathbb{G}'_A)$ detects whether S_A is a fiber. See also Corollary 9.16 on p. 149.

Theorem 5.11. *Let $D(K)$ be any link diagram, and let S_A be the spanning surface determined by the all-A state of this diagram. Then the following are equivalent:*

(1) *The reduced graph $\mathbb{G}'_A$ is a tree.*
(2) $S^3 \setminus K$ *fibers over S^1, with fiber S_A.*
(3) $M_A = S^3 \backslash\backslash S_A$ *is an I-bundle over S_A.*

We would like to emphasize that the theorem applies to *all* diagrams. It turns out that each of (1), (2), and (3) implies that D is connected and A-adequate. The point of including condition (3) is that S_A is never a *semi-fiber*: that is, S_A cannot be a non-orientable surface that lifts to a fiber in a double cover of $S^3 \setminus K$.

Proof. For (1) $\Rightarrow$ (2), suppose that $\mathbb{G}'_A$ is a tree. Then D must be connected because $\mathbb{G}'_A$ is connected. Also, since $\mathbb{G}'_A$ contains no loops, $\mathbb{G}_A$ must contain no 1-edge loops, hence D is A-adequate. In particular, we have a polyhedral decomposition of $M_A = S^3 \backslash\backslash S_A$, and all the results of the previous chapters apply to this polyhedral decomposition.

Since $\mathbb{G}'_A$ is a tree, Lemma 5.8(4) implies that every polyhedron of the polyhedral decomposition is a prism, and every white face is a bigon. Observe that a prism is an I-bundle over its base polygon, and a white bigon face is also an I-bundle with the same product structure. Thus the I-bundle structures of the individual polyhedra can be glued along the bigon faces to obtain an I-bundle structure on all of M_A.

Finally, since $\mathbb{G}'_A$ is a tree, it is bipartite, hence $\mathbb{G}_A$ is also bipartite. Thus, by Lemma 2.3 on p. 18, S_A is orientable. Since S_A is an orientable surface whose complement is an I-bundle, it must be a fiber in a fibration over S^1.

The implication (2) $\Rightarrow$ (3) is trivial.

For (3) $\Rightarrow$ (1), suppose that M_A is an I-bundle over S_A. Thus, in particular, S_A is connected, hence D is connected. Also, S_A must be essential in $S^3 \setminus K$. Thus, by Theorem 3.19, D is A-adequate, and all of our polyhedral techniques apply.

All white faces of the polyhedral decomposition are contained in M_A. Thus, by Lemma 4.17 (Product rectangle in white face), each white face is a product $\alpha \times I$, where $\alpha \times \{0, 1\}$ are ideal edges. In other words, every white face is a bigon. Thus, by Lemma 5.8, $\mathbb{G}'_A$ is a tree.   □

Remark 5.12. We have seen in Lemma 2.21 that each polyhedral region corresponds to an alternating link diagram, whose all-A surface is a checkerboard surface. Ozawa has observed that the state surface S_A is a Murasugi sum of these individual checkerboard surfaces [76]; this was the basis of his proof that S_A is essential (Theorem 3.19). Now, a theorem of Gabai [39, 40] states that the Murasugi sum of several surfaces is a fiber if and only if the individual summands are fibers. Thus an alternate proof of Theorem 5.11 would argue by induction: here, the base case is that of prime, alternating diagrams, and Gabai's result gives the inductive step.

Restricted to prime, alternating diagrams, Theorem 5.11 says that the checkerboard surface S_A is a fiber if and only if $D(K)$ is a negative 2-braid. (We are using the convention that positive braid generators are as depicted in Fig. 9.1 on p. 141.) This special case follows quickly from a theorem of Adams [2, Theorem 1.9], and can also be proved by applying Lemma 4.17 to Menasco's polyhedral decomposition of alternating link complements [64].

In fact, this line of argument extends to give a version of Theorem 5.11 for state surfaces of σ-homogeneous states (i.e., Theorem 5.21 below). See the recent paper by Futer [29] for a proof from this point of view.

5.4 Computing the Guts

To compute the guts of $M_A = S^3 \backslash\backslash S_A$, it suffices to take the spanning sets of the previous section, count the EPDs in the spanning sets, and also count how many prisms will occur in the complement of these disks. The counts work as follows.

Proposition 5.13. *Every non-trivial component of the characteristic submanifold of M_A is spanned by a collection $E_l \cup E_c$ of essential product disks, such that*

(1) *The disks of E_l are embedded in lower polyhedra, and $||E_l|| = e_A - e'_A + n_{\text{sep}}$, where n_{sep} is the number of separating edges in $\mathbb{G}'_A$.*
(2) *The disks of E_c are embedded in the upper polyhedron. All these disks are complex. Furthermore, no disk in E_c parabolically compresses to bigon faces and other disks in E_c.*
(3) *After the characteristic submanifold is cut along $E_l \cup E_c$, the total number of prism pieces will be $n_{\text{sep}} + \chi_+(\mathbb{G}'_A)$, where n_{sep} is the number of separating edges in $\mathbb{G}'_A$ and $\chi_+(\mathbb{G}'_A) = \max\{0, \chi(\mathbb{G}'_A)\}$ equals 1 if $\mathbb{G}'_A$ is a tree and 0 otherwise.*

Proof. By Theorem 4.4, every non-trivial component of the characteristic submanifold is spanned by EPDs in individual polyhedra. We have constructed spanning sets for the individual polyhedra in Lemmas 5.6 and 5.8; these sets are denoted E_l in the lower polyhedra and $E_s \cup E_c$ in the upper polyhedron. Note that by construction, every white bigon face in the polyhedral decomposition has a disk in E_l parallel to it, as well as a disk in E_s parallel to it. We do not need both of these parallel disks to span the characteristic submanifold. Thus we may discard E_s, and conclude that $E_l \cup E_c$ spans the characteristic submanifold.

Conclusion (1), which counts the cardinality of E_l, is a restatement of Lemma 5.7. Conclusion (2) is a restatement of Lemma 5.8(2).

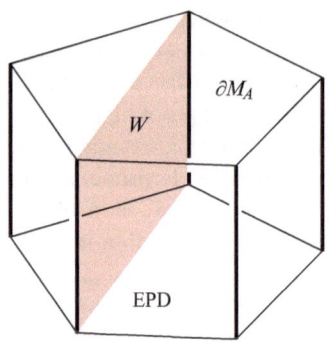

Fig. 5.3 A prism R, whose lateral faces are EPDs in the spanning set. The parabolic locus is in *bold*. Any *white* face W that intersects R must respect the product structure of R, hence is a bigon face

It remains to count the prism components cut off by $E_l \cup E_c$. Recall that every white bigon face in the polyhedral decomposition has a disk in E_l parallel to it, as well as a disk in E_s parallel to it. Thus every prism cut off by $E_l \cup E_c$ is isotopic (through white bigon faces) to a prism cut off by $E_l \cup (E_c \cup E_s)$. By Lemma 5.7, the number of these prisms in the lower polyhedra is equal to n_{sep}. By Lemma 5.8(4), the number of these prisms in the upper polyhedron is 0 or 1, and is equal to $\chi_+(\mathbb{G}'_A)$. Thus the proof will be complete once we show that every prism cut off by $E_l \cup E_c$ is isotopic into a single polyhedron.

Let R be a prism over an n-gon, cut off by $E_l \cup E_c$. Suppose that there is a white face W of the polyhedral decomposition that intersects R (otherwise we are done). By Lemma 4.17 (Product rectangle in white face), each component of $R \cap W$ is a product rectangle $\alpha \times I$, whose top and bottom sides $\alpha \times \{0, 1\}$ are sub-arcs of edges of W. But by construction, each lateral face of R is an EPD belonging to $E_l \cup E_c$, hence lies in a single polyhedron and is disjoint from W. Thus $\alpha \times \{0, 1\}$ must be disjoint from the lateral EPDs, and must run from the parabolic locus to the parabolic locus. In other words, $\alpha \times I$ fills up the entirety of the white face W, hence W is a bigon. See Fig. 5.3.

Recall that by Lemma 5.7, every simple disk in the lower polyhedra belongs to E_l. Thus W is parallel to a disk of E_l, hence to a lateral face of the prism R. By isotoping R through this white face W, we move a lateral face of R from a disk of E_l to a parallel disk of E_s, while removing a component of intersection with the white faces. Continuing inductively in this fashion, we conclude that if R was not already in a single polyhedron, it can be isotoped into the top polyhedron. Thus R was already accounted for in the count of $n_{\mathrm{sep}} + \chi_+(\mathbb{G}'_A)$ prisms, and the proof is complete.   □

We can now prove the main theorem.

Theorem 5.14. *Let $D(K)$ be an A-adequate diagram, and let S_A be the essential spanning surface determined by this diagram. Then*

$$\chi_-(\mathrm{guts}(S^3 \backslash\backslash S_A)) = \chi_-(\mathbb{G}'_A) - \|E_c\|,$$

where $\chi_-(\cdot)$ is the negative Euler characteristic as in Definition 1.5, and where $||E_c||$ is the smallest number of complex disks required to span the I-bundle of the upper polyhedron, as in Definition 5.9.

In particular, if every essential product disk in the upper polyhedron is simple or semi-simple, then

$$\chi_-(\text{guts}(S^3\backslash\backslash S_A)) = \chi_-(\mathbb{G}'_A).$$

Proof. Recall that the graph $\mathbb{G}_A$ embeds as a spine for the surface S_A. Thus, by Alexander duality, $M_A = S^3\backslash\backslash S_A$ has Euler characteristic

$$\chi(M_A) = \chi(S_A) = \chi(\mathbb{G}_A). \tag{5.1}$$

Recall that $M_A = \text{guts}(M_A) \cup CS(M_A)$, where $CS(M_A)$ is the characteristic submanifold of M_A, and the intersection $\text{guts}(M_A) \cap CS(M_A)$ consists of annuli. Thus their Euler characteristics sum to the Euler characteristic of M_A. Furthermore, by Lemma 4.1, all the trivial components of the I-bundle are solid tori glued along annuli, which do not contribute to the Euler characteristic count. Therefore, if B denotes the maximal I-bundle in the characteristic submanifold,

$$\chi(\mathbb{G}_A) = \chi(M_A) = \chi(\text{guts}(M_A)) + \chi(B). \tag{5.2}$$

By Proposition 5.13, the maximal I-bundle B is spanned by a collection $E_l \cup E_c$ of essential product disks. Notice that cutting B along a disk increases its Euler characteristic by 1. By Definition 4.3, we know that $B\backslash\backslash(E_l \cup E_c)$ consists of solid tori (Euler characteristic 0) and prisms (Euler characteristic 1). Thus, by Proposition 5.13,

$$
\begin{aligned}
\chi(B) &= -||E_l \cup E_c|| & &+ \text{(number of prisms)} \\
&= -(e_A - e'_A) - n_{\text{sep}} - ||E_c|| & &+ (n_{\text{sep}} + \chi_+(\mathbb{G}'_A)) \\
&= \chi(\mathbb{G}_A) - \chi(\mathbb{G}'_A) - ||E_c|| & &+ \chi_+(\mathbb{G}'_A) \\
&= \chi(\mathbb{G}_A) + \chi_-(\mathbb{G}'_A) - ||E_c||.
\end{aligned}
\tag{5.3}
$$

Since every component of $\text{guts}(M_A)$ is bounded by a hyperbolic surface, we have $\chi_-(\text{guts}(M_A)) = -\chi(\text{guts}(M_A))$. Thus plugging the conclusion of (5.3) into (5.2) gives

$$\chi_-(\text{guts}(M_A)) = -\chi(\text{guts}(M_A)) = \chi_-(\mathbb{G}'_A) - ||E_c||,$$

which completes the proof. $\square$

Remark 5.15. The reliance on the notation $\chi_-(\cdot)$ in Theorem 5.14 is only necessary in the special case when $\mathbb{G}'_A$ is a tree and $\text{guts}(M_A)$ is empty. On the other hand, when $\text{guts}(M_A) \neq \emptyset$, every component of it will have negative Euler characteristic. Thus, when $\text{guts}(M_A) \neq \emptyset$, the conclusion of the theorem can be rephrased as

$$\chi(\text{guts}(M_A)) = \chi(\mathbb{G}'_A) - ||E_c|| < 0.$$

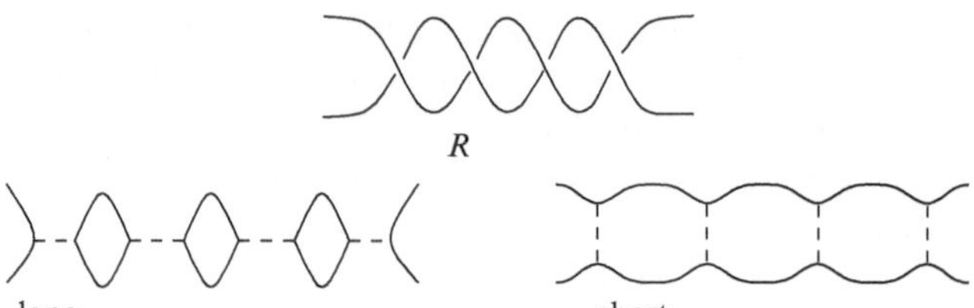

Fig. 5.4 Resolutions of a twist region R. This twist region is an A-region, because the all-A resolution is short

5.5 Modifications of the Diagram

In this section we use Theorem 5.14 to study the effect on the guts of two well known link diagrammatic moves: adding/removing crossings to a twist region, and taking planar cables. Lemma 5.17 characterizes the effect on the guts of adding/removing crossings to a twist region. Planar cables, which are of particular significance to us since they are used in the calculation of the colored Jones polynomials [60], are discussed in Corollary 5.20.

Definition 5.16. Let R be a twist region of the diagram D, and suppose that R contains $c_R > 1$ crossings. Consider the all-A and all-B resolutions of R. One of the graphs associated to D, say $\mathbb{G}_B$, will inherit $c_R - 1$ vertices from the $c_R - 1$ bigons contained in R. We say that this is the *long resolution* of the twist region R. The other graph, say $\mathbb{G}_A$, contains c_R parallel edges (only one of which survives in $\mathbb{G}'_A$). This is the *short resolution* of R. See Fig. 5.4.

We say that the twist region R is an *A-region* if the all-A resolution is the short resolution of R. In other words, R is an A-region if it contributes exactly one edge to $\mathbb{G}'_A$.

Lemma 5.17. *Let D be an A-adequate link diagram, with spanning surface $S_A(D)$ and the associated prime polyhedral decomposition of $S^3 \backslash\backslash S_A(D)$. Let $\widehat{D}$ be the A-adequate diagram obtained by removing one crossing in an A-region of D. (Note that this operation very likely changes the link type.)*

Then the effect of removing one crossing from an A-region is as follows:

(1) *The reduced graphs $\mathbb{G}'_A(D)$ and $\mathbb{G}'_A(\widehat{D})$ are isomorphic.*
(2) *In the upper polyhedra of the respective diagrams, the spanning sets $E_c(D)$ and $E_c(\widehat{D})$ have the same cardinality.*
(3) *The complements of the spanning surfaces $S_A(D)$ and $S_A(\widehat{D})$ have the same guts:*
$$\chi_- \operatorname{guts}(S^3 \backslash\backslash S_A(D)) = \chi_- \operatorname{guts}(S^3 \backslash\backslash S_A(\widehat{D})).$$

Proof. Following Definition 5.16, let R be a twist region of the diagram D which has at least two crossings, and in which the all-A resolution is short. Then, removing one crossing from twist region R amounts to removing one segment from the graph

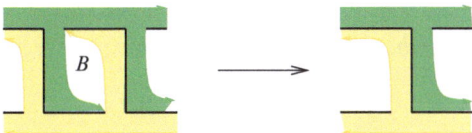

Fig. 5.5 The effect of removing a crossing from an A-region of D on the graph H_A and the polyhedral decomposition. In the *upper* polyhedron, a bigon face B between two *shaded* faces becomes collapsed to a single ideal vertex

$H_A(D)$. All the state circles are unaffected, and the other segments of H_A are also unaffected. See Fig. 5.5.

Recall that the vertices of $\mathbb{G}_A$ are the state circles of H_A, and the edges of $\mathbb{G}_A$ are the segments of H_A. Thus the graphs $\mathbb{G}_A(D)$ and $\mathbb{G}_A(\widehat{D})$ have exactly the same vertex set, with $\mathbb{G}_A(D)$ having one more edge in the short resolution of the twist region R. Because duplicate edges of $\mathbb{G}_A$ are identified together in $\mathbb{G}'_A$, the two reduced graphs $\mathbb{G}'_A(D)$ and $\mathbb{G}'_A(\widehat{D})$ are isomorphic, proving (1).

Now, consider the effect of removing a crossing from R on the polyhedral decomposition. A bigon in the twist region R corresponds to a white bigon face of the upper polyhedron in the polyhedral decomposition of D. Let F and G be the two shaded faces that are adjacent to this bigon B. As Fig. 5.5 shows, removing one crossing from R amounts to collapsing the bigon face B to a single ideal vertex.

Next, consider the essential product disks through faces F and G that form part of the spanning set $E_s(D) \cup E_c(D)$. By Lemma 5.8(1), the simple disk parallel to bigon B is part of the spanning set $E_s(D)$. Furthermore, by Lemma 5.5, all other PITOS disks through faces F and G remain PITOS if we collapse B to a single ideal vertex.

Recall that in the proof of Lemma 5.8, we considered two cases. If F and G are the only shaded faces in the upper polyhedron, then $E_c = \emptyset$. This will remain true after we remove one bigon face. Alternately, if F and G are not the only shaded faces, then the contribution of these shaded faces to $E_s(D) \cup E_c(D)$ consists of all PITOS disks through F and G. The property that a PITOS disk is complex will not change as we collapse the bigon B. Thus $||E_c(D)|| = ||E_c(\widehat{D})||$, proving (2).

Finally, (3) follows immediately from (1), (2), and Theorem 5.14. $\qquad\square$

Remark 5.18. For alternating diagrams, Lackenby observed that there is actually a homeomorphism from $\mathrm{guts}(S^3 \backslash\backslash S_A(D))$ to $\mathrm{guts}(S^3 \backslash\backslash S_A(\widehat{D}))$, which carries parabolic locus to parabolic locus. See [58, P. 215]. This statement holds in complete generality, including in our setting. However, we will only need the equality of Euler characteristics in Lemma 5.17(3).

By combining Theorem 5.14 and Lemma 5.17 with Theorem 6.4 on p. 93 (which will be proved in the next chapter), we obtain the following corollary.

Corollary 5.19. *Suppose that $D(K)$ is a prime, A-adequate diagram, such that for each 2-edge loop in $\mathbb{G}_A$ the edges belong to the same twist region of $D(K)$. Then*

$$\chi_-(\mathrm{guts}(M_A)) = \chi_-(\mathbb{G}'_A).$$

Proof. Let D be as in the statement of the corollary, and let $\widehat{D}$ be the diagram that results from removing *all* bigons in the A-regions of the diagram D. Applying Lemma 5.17 inductively, we conclude that this removal of bigons does not affect either the reduced graph $\mathbb{G}'_A$ or spanning set E_c. Also, since every 2-edge loop of $\mathbb{G}_A(D)$ belongs to a single twist region, the removal of bigons also removes all 2-edge loops. Thus $\mathbb{G}'_A(D) = \mathbb{G}'_A(\widehat{D}) = \mathbb{G}_A(\widehat{D})$.

By Theorem 6.4, every essential product disk in the upper polyhedron of $\widehat{D}$ must run over tentacles adjacent to the segments of a 2-edge loop of $\mathbb{G}_A(\widehat{D})$. But by construction, there are no 2-edge loops in $\mathbb{G}_A(\widehat{D})$. Thus, by Lemma 5.17, $E_c(D) = E_c(\widehat{D}) = \emptyset$. Therefore, according to the formula of Theorem 5.14, $\chi_-(\mathrm{guts}(M_A)) = \chi_-(\mathbb{G}'_A)$.   □

Given a diagram $D = D(K)$ of a link K, and a number $n \in \mathbb{N}$, let D^n denote the n-cabling of D using the blackboard framing. If D is A-adequate then D^n is A-adequate for all $n \in \mathbb{N}$. Furthermore, the Euler characteristic of the reduced all-A graph corresponding to D^n, is the same as that of the reduced all-A graph corresponding to D. That is,

$$\chi(\mathbb{G}'_A(D^n)) = \chi(\mathbb{G}'_A(D)),$$

for all $n \geq 1$ [60, Chap. 5]. We have the following:

Corollary 5.20. *Let $D := D(K)$ be an A-adequate diagram, of a link K. Let D^n denote the n-cabling of D using the blackboard framing, and let S^n_A be the all-A state surface determined by D^n. Then*

$$\chi_-(\mathrm{guts}(S^3 \backslash\backslash S^n_A)) + ||E_c(D^n)|| = \chi_-(\mathbb{G}'_A(D)),$$

for every $n \geq 1$. Here $\chi_-(\cdot)$, $||\cdot||$ and $E_c(D^n)$ are the quantities of the statement of Theorem 5.14 corresponding to D^n.

Proof. By Theorem 5.14, we have

$$\chi_-(\mathrm{guts}(S^3 \backslash\backslash S^n_A)) + ||E_c(D^n)|| = \chi_-(\mathbb{G}'_A(D^n)).$$

Since $\chi(\mathbb{G}'_A(D^n)) = \chi(\mathbb{G}'_A(D))$, for all $n \geq 1$, the result follows.   □

It is worth observing that by Corollary 5.20 and Theorem 5.14,

$$\chi_-(\mathrm{guts}(S^3 \backslash\backslash S^n_A)) + ||E_c(D^n)|| = \chi_-(\mathrm{guts}(S^3 \backslash\backslash S_A)) + ||E_c(D)||$$
$$= \chi_-(\mathbb{G}'_A(D)),$$

for every $n \geq 1$. Thus the left-hand side is independent of n. It is worth asking whether the summands $\chi_-(\text{guts}(S^3 \backslash\backslash S_A^n))$ and $||E_c(D^n)||$ are also independent of n; see Question 10.5 in Chap. 10.

In fact, by Lemma 9.14 on p. 149, the quantity $\chi_-(\mathbb{G}'_A(D))$ is actually an invariant of the link K; it is independent of the A-adequate diagram. However, Example 5.3 on p. 74 demonstrates that $||E_c(D)||$ (and thus $\chi_-(\text{guts}(S^3 \backslash\backslash S_A))$) does, in general, depend on the diagram used: Fig. 5.1 shows a diagram D with $||E_c(D)|| \neq 0$, while Fig. 5.2 shows a different diagram D' of the same link with $||E_c(D')|| = 0$. We will revisit this discussion in Chap. 10.

5.6 The σ-Adequate, σ-Homogeneous Setting

The results of this chapter extend immediately to σ-adequate, σ-homogeneous states, using only the fact that the polyhedral decomposition in this case cuts M_σ into checkerboard polyhedra (Theorem 3.23). This is because the proofs in this section use only normal surface theory specific to checkerboard polyhedra, and nothing dependent on tentacles or choice of resolution.

In particular, Lemma 5.1 holds, and its proof needs no change, using the fact that lower polyhedra still correspond to checkerboard polyhedra of alternating links. Definitions 5.2 and 5.4, as well as Lemma 5.5, are all stated (and proved) for any checkerboard colored ideal polyhedron. Lemmas 5.6 and 5.7 concern only lower polyhedra, which we know correspond to ideal polyhedra of alternating links in the σ-homogeneous case. Hence their proofs will hold in this general setting. Lemmas 5.8 and 5.10, concerning upper polyhedra, use only properties of checkerboard ideal polyhedra, hence these lemmas still hold if we replace $\mathbb{G}'_A$ with G'_σ. Similarly, Theorem 5.11 will immediately generalize to the σ-adequate, σ-homogeneous setting. The proof uses Lemma 5.8 and Proposition 4.17 (Product rectangle in white face), and we have seen that these results hold in the σ-adequate, σ-homogeneous case. Thus the proof of Theorem 5.11 applies verbatim to give the following general result.

Theorem 5.21. *Let $D(K)$ be a link diagram, and let S_σ be the state surface of a homogeneous state σ. Then the following are equivalent.*

(1) *The reduced graph G'_σ is a tree.*
(2) *$S^3 \setminus K$ fibers over S^1 with fiber S_σ.*
(3) *$M_\sigma = S^3 \backslash\backslash S_\sigma$ is an I-bundle over S_σ.* □

In particular, Theorem 5.21 implies the classical result, due to Stallings [89], that homogeneous closed braids are fibered, with fiber the Seifert surface S_σ associated to the Seifert state σ.

The results of Sect. 5.4 will also extend immediately to the σ-adequate, σ-homogeneous setting. In particular, every non-trivial component of the characteristic submanifold of M_σ is spanned by a collection $E_l \cup E_c$ of EPDs, with the properties of Proposition 5.13, with σ replacing A in the appropriate places. In the proof of Proposition 5.13, one would need to use Theorem 4.22 in place of Theorem 4.4. Theorem 5.14 also generalizes to this setting, and we obtain

$$\chi_-(\text{guts}(S^3 \backslash\backslash S_\sigma)) = \chi_-(\mathbb{G}'_\sigma) - ||E_c||.$$

As for Sect. 5.5, in the case of a σ-adequate, σ-homogeneous diagram, analogous to an A-region, we define a twist region R to be a σ-region if its σ-resolution gives the short resolution of R. In other words, R contributes exactly one edge to $\mathbb{G}'_\sigma$. With this modification, Lemma 5.17 will hold, with the same proof, replacing A-adequate with σ-adequate, σ-homogeneous in the statement, as well as S_A with S_σ, A-region with σ-region, and $\mathbb{G}'_A$ with $\mathbb{G}'_\sigma$.

However, this is where we stop. Corollary 5.19 requires results from Chap. 6, which we have not analyzed in the σ-adequate, σ-homogeneous case. It is entirely possible that the results of Chap. 6 will generalize, but we leave this analysis for a future time.

Chapter 6
Recognizing Essential Product Disks

Theorem 5.14 reduces the problem of computing the Euler characteristic of the guts of M_A to counting how many complex EPDs are required to span the I-bundle of the upper polyhedron. Our purpose in this chapter is to recognize such EPDs from the structure of the all-A state graph $\mathbb{G}_A$. The main result is Theorem 6.4, which describes the basic building blocks for such EPDs. Each corresponds to a 2-edge loop of the graph $\mathbb{G}_A$.

The proofs of this chapter require detailed tentacle chasing arguments, and some are quite technical. To assist the reader, we break the proof of Theorem 6.4 into four steps, and keep a running outline of what has been accomplished, and what still needs to be accomplished. The tentacle chasing does pay off, for by the end of the chapter we obtain a mapping from any EPD to one of only seven possible sub-graphs of H_A. By investigating the occurrence of such subgraphs, we are able to count complex EPDs in large classes of link complements. Two such classes are studied in detail in Chaps. 7 (links with diagrams without non-prime arcs) and 8 (Montesinos links). Together with the results of Chap. 5, these give applications to guts, volumes, and coefficients of the colored Jones polynomials.

6.1 2-Edge Loops and Essential Product Disks

To find essential product disks in the upper polyhedron, we will convert any EPD into a normal square and use machinery developed in Chap. 4.

Lemma 6.1 (EPD to oriented square). *Let D be a prime, A-adequate diagram of a link in S^3, with prime polyhedral decomposition of $M_A = S^3 \backslash\backslash S_A$. Suppose there is an EPD embedded in M_A in the upper polyhedron. Then the boundary of the EPD can be pulled off the ideal vertices to give a normal square in the polyhedron with the following properties.*

D. Futer et al., *Guts of Surfaces and the Colored Jones Polynomial*, Lecture Notes in Mathematics 2069, DOI 10.1007/978-3-642-33302-6_6,
© Springer-Verlag Berlin Heidelberg 2013

(1) *Two opposite edges of the square run through shaded faces, which we label green and orange.*[1]
(2) *The other two opposite edges run through white faces, each cutting off a single vertex of the white face.*
(3) *The single vertex of the white face, cut off by the white edge, is a triangle, oriented such that in counter-clockwise order, the edges of the triangle are colored orange–green–white.*

With this convention, the two white edges of the normal square cannot lie on the same white face of the polyhedron.

Proof. The EPD runs through two shaded faces, green and orange, and two ideal vertices. Any ideal vertex meets two white faces. Thus we may push an arc running over an ideal vertex slightly off the vertex to run through one of the adjacent white faces instead. Note that there are two choices of white face into which we may push the arc, giving oppositely oriented triangles. For each vertex, we choose to push in the direction that gives the triangle oriented as in the statement of the lemma.

Finally, to see that the two white edges of the normal square do not lie on the same white face, we argue by contradiction. Suppose the two white edges do lie on the same white face. The white faces are simply connected, so we may run an arc from one to the other through the white face. Since the shaded faces are simply connected, we may run an arc through the green face meeting the white edges of the square at their boundaries. Then the union of these two arcs gives a closed curve which separates the two ideal vertices. This contradicts Proposition 3.18 (No normal bigons).   □

As in Chap. 4, call the arcs of the normal square which lie on white faces β_W and β_V.

We will use Lemma 6.1 (EPD to oriented square) to prove Theorem 6.4, which is the main result of this chapter. Before we state the theorem we need two definitions. For the first, note that the portion of a tentacle adjacent to a segment has a natural product structure, homeomorphic to the product of the segment and an interval. In particular, the center point p of the segment defines a line $p \times I$ running across the tentacle.

Definition 6.2. We say that an arc through the tentacle runs *adjacent* to a segment s if it runs transversely exactly once through the line $p \times I$, where p is the center of the segment, and the portion of the tentacle adjacent to the segment is homeomorphic to $s \times I$.

Definition 6.3. Recall that an ideal vertex in the upper polyhedron is described on the graph H_A by a connected component of the knot, between two undercrossings. Such vertices will be right-down staircases, containing zero or more segments of H_A. A *zig-zag* is defined to be one of these ideal vertices.

---

[1]Note: For grayscale versions of this chapter, the figures will show green faces as darker gray, orange faces as lighter gray.

Theorem 6.4. *Let $D(K)$ be a prime, A-adequate diagram of a link K in S^3, with prime polyhedral decomposition of $M_A = S^3 \backslash\backslash S_A$. Suppose there is an essential product disk embedded in M_A in the upper polyhedron, with associated normal square of Lemma 6.1 (EPD to oriented square). Then there is a 2-edge loop in $\mathbb{G}_A$ so that the normal square runs over tentacles adjacent to segments of the 2-edge loop.*

Moreover, the normal square has one of the types $\mathscr{A}$–$\mathscr{G}$ shown in Fig. 6.1.

Before we proceed with the proof of Theorem 6.4, which will occupy the remainder of this chapter, we describe the properties and features of the types $\mathscr{A}$ through $\mathscr{G}$ in some detail. We want to emphasize that the colors have been selected so that colors and orientations at vertices must be exactly as described, or exactly as shown in Fig. 6.1. This is a consequence of the choice of orientation in Lemma 6.1 (EPD to oriented square).

$\mathscr{A}$. The square runs through distinct shaded faces adjacent to the two segments. One vertex of the EPD is a zig-zag (possibly without segments) with one end on one of the state circles met by the two segments of the loop, and the arcs of the normal square both run adjacent to this zig-zag along its length, meeting at a white face at its end.

$\mathscr{B}$. The square runs through the same (orange) shaded face adjacent to the two segments. One vertex of the EPD is a zig-zag with one end on one of the state circles met by the two segments as before, with the arcs of the normal square running adjacent to this zig-zag along its length, meeting at a white face at its end.

$\mathscr{C}$. The boundary of the EPD runs through distinct shaded faces adjacent to two segments, as in type $\mathscr{A}$ above, and as in that case the vertex has an end on one of the two state circles met by the two segments. However, in this case the vertex is on the opposite side of the 2-edge loop, and so the boundary of the EPD at the vertex does not run adjacent to the zig-zag of the vertex, but immediately runs into a white face.

$\mathscr{D}$. The boundary of the EPD runs through the same shaded face adjacent to the two segments, as in type $\mathscr{B}$ above, and meets a vertex on one of the two state circles, but the vertex is on the opposite side as that in type $\mathscr{B}$. With colors and orientations chosen, this forces the square to run through two green tentacles, whereas in type $\mathscr{B}$ it must run through two orange tentacles.

$\mathscr{E}$. The boundary of the EPD runs through distinct shaded faces adjacent to two segments, which are separated from one of the vertices by a non-prime arc.
    After running downstream adjacent to one of the segments (inside non-prime arc in figure shown), the boundary of the EPD immediately crosses at least one non-prime arc with endpoints on the same state circle as the segment. On the other side of these separating non-prime arcs, the boundary of the EPD runs directly to one of the vertices.

$\mathscr{F}$. Identical to type $\mathscr{E}$, only the 2-edge loop runs through two green faces rather than distinct colors.

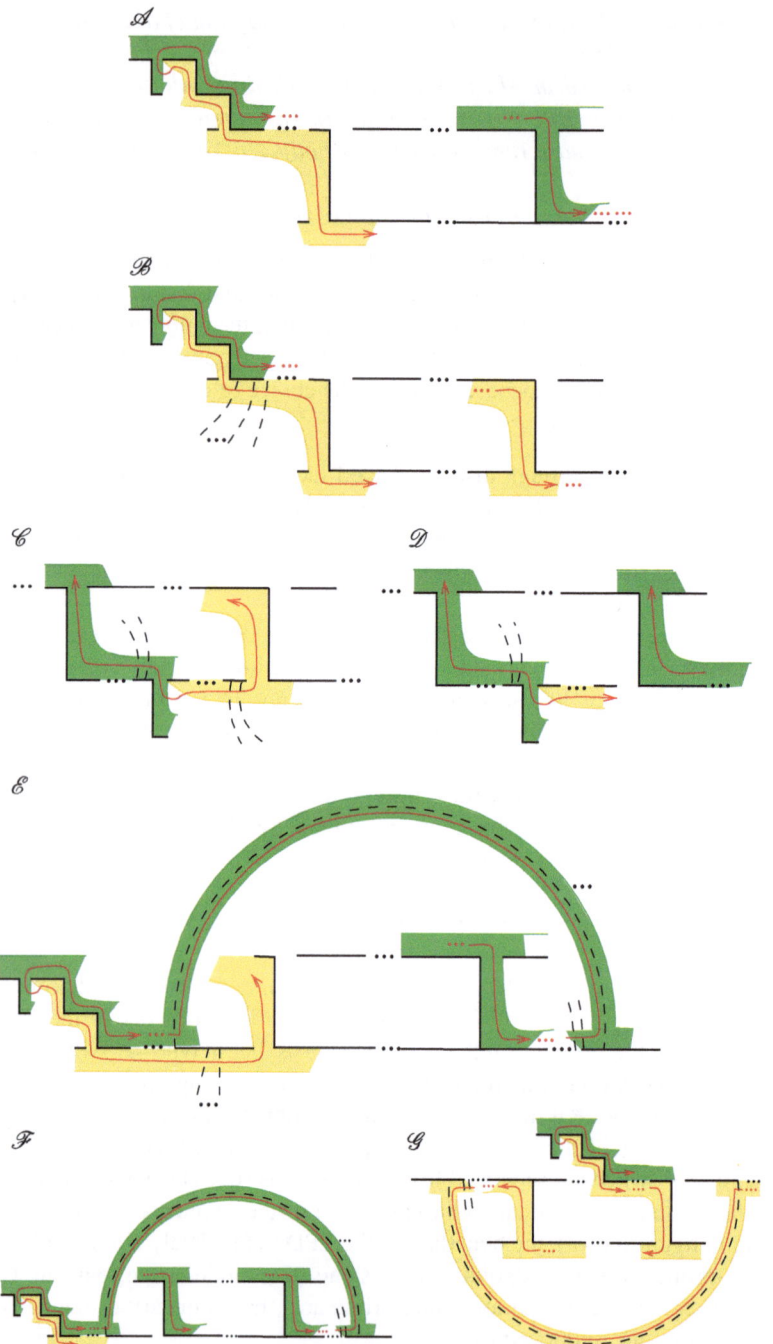

Fig. 6.1 Building blocks of EPDs in top polyhedron (Note: For grayscale versions of this monograph, *green* faces will appear as *dark gray*, orange faces as *lighter gray*)

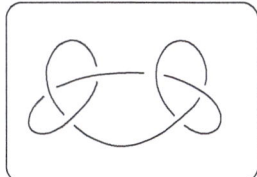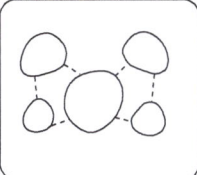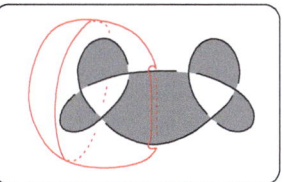

Fig. 6.2 *Left*: The connect sum of two left-handed trefoils. *Middle*: The all-A state. Note there are no 2-edge loops. *Right*: The state surface S_A with EPD shown in *red*

$\mathcal{G}$. Similar to type $\mathcal{F}$, only the 2-edge loop runs through two orange faces. Because the faces are orange, the zig-zag vertex (which still may contain no segments), is adjacent to the opposite side of a state circle meeting both segments of the 2-edge loop.

Remark 6.5. In the statement of Theorem 6.4, we require the diagram to be prime, as in Definition 1.7. We have not used the hypothesis of prime diagrams until now, but it will be crucial going forward. In fact, Theorem 6.4 does not hold for diagrams that are not prime. For example, the connected sum of two left-handed trefoils is not prime, and its all-A state graph $\mathbb{G}_A$ has no 2-edge loops. See Fig. 6.2. But, if Σ is the sphere along which we performed the connect sum, then $\Sigma \backslash\backslash S_A$ is an essential product disk in $S^3 \backslash\backslash S_A$. One may check that this EPD is isotopic into the upper polyhedron (indeed, by Lemma 5.1 on p. 73, there are no EPDs in the lower polyhedra).

Before embarking on the proof of Theorem 6.4, we record the following corollary of the theorem, which removes the dependence on the pulling-off procedure of Lemma 6.1.

Corollary 6.6. *Let $D(K)$ be a prime, A-adequate diagram of a link K in S^3, with prime polyhedral decomposition of $M_A = S^3 \backslash\backslash S_A$. Let E be an essential product disk embedded in the upper polyhedron of M_A. Then ∂E runs over tentacles adjacent to segments of a 2-edge loop, of one of the types $\mathscr{A}$ through $\mathscr{G}$ shown in Fig. 6.1.*

Proof. The essential product disk E may be pulled off the ideal vertices of the upper polyhedron P, as in Lemma 6.1 (EPD to oriented square). By Theorem 6.4, the resulting normal square S must run over a 2-edge loop, as in Fig. 6.1. We may recover E from S by pulling the segments of ∂S in the white faces back onto the ideal vertices of the polyhedron P.

Recall, from Definition 6.3, that an ideal vertex of P is seen as a zig-zag (right-down staircase) on the graph H_A. Thus, after we pull ∂S back onto the ideal vertices, the disk E will cross from one shaded face into the other at some point of the zigzag. Performing this operation in panels $\mathscr{A}$ through $\mathscr{G}$ in Fig. 6.1, we see that ∂E still runs over tentacles adjacent to segments of a 2-edge loop. □

6.2   Outline and First Step of Proof

As mentioned above, the proof of Theorem 6.4 requires a significant amount of tentacle chasing, which is done in the remainder of this chapter. The reader who wishes to avoid tentacle chasing for now may move on to Chap. 8 and continue reading from there. Chapter 7, which is independent from Chaps. 8–10, will also involve tentacle chasing, and requires results from Sect. 6.3 below.

In addition to tentacle chasing, the proof of Theorem 6.4 requires the analysis of several cases. In each case, we show either there is a 2-edge loop of the proper form, or that we can further restrict the diagram. Thus as the proof progresses, we are left with more and more restrictions on the diagram, until we analyze a handful of special cases to finish the proof of the theorem. The proof follows four basic steps, which we summarize as follows.

1. Prove Theorem 6.4 holds if β_V and β_W lie in the same polyhedral region, where recall β_V and β_W denote the sides of the normal square which lie on white faces.
2. If β_V and β_W are not in the same polyhedral region, then prove the theorem holds, or β_V and β_W run through a portion of the diagram of one of two particular forms near β_V, β_W. These are illustrated in Fig. 6.5 on p. 99.
3. Prove the theorem holds, or the normal square runs along both sides of a zig-zag—one of the vertices of the EPD—and then into a non-prime arc separating β_V and β_W.
4. Analyze behavior inside a first separating non-prime arc.

Step 1 of the proof is treated in the next lemma, which shows that Theorem 6.4 holds if β_V and β_W lie in the same polyhedral region.

Lemma 6.7 (Step 1). *Suppose we have a prime, A-adequate diagram with prime polyhedral decomposition, and an EPD intersecting white faces V and W in arcs β_V and β_W, respectively. If V and W are in the same polyhedral region, then there is a 2-edge loop of type $\mathscr{B}$. In particular, Theorem 6.4 holds in this case.*

Proof. Apply the clockwise map. Lemma 4.8 implies that we may join the images of β_V and β_W into a square S' in the lower polyhedron. Because each of β_V, β_W cuts off a single ideal vertex in the upper polyhedron, each will cut off a single ideal vertex in the lower polyhedron, with the same orientation as in the upper polyhedron, and thus this new square S' either is inessential, or can be isotoped to an essential product disk. We will treat the two cases separately.

Case 1: S' is isotoped to an essential product disk in a lower polyhedron. Then Lemma 5.1 implies that the disk runs over two segments of H_A corresponding to a 2-edge loop in the lower polyhedron. This loop must come from a 2-edge loop in the upper polyhedron. In the lower polyhedron, the image of the arc β_W must be the arc on the left of Fig. 6.3, adjacent to the segment on the left. Similarly, the image of β_V must be adjacent to the segment on the right. The preimages of these arcs are shown on the right of Fig. 6.3. Note that the dashed lines on the portion of the

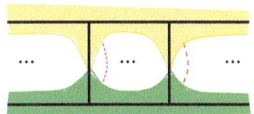  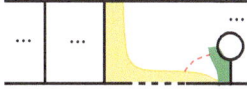

Fig. 6.3 *Left*: *Red* (*dotted* and *dashed*) arcs show images of β_W under the clockwise map. *Right*: preimages of arcs on *left*

graph of H_A on the right are to indicate that the ideal edge may run over non-prime switches between its head and tail, which will not affect the argument.

For both vertices, there is an (orange) tentacle shown whose tail meets the vertex, which runs upstream adjacent to a segment. We will show that σ_1 and σ_2 run upstream through these tentacles, and so the 2-edge loop is of type $\mathscr{B}$ of Theorem 6.4.

Suppose first that σ_1 or σ_2 crosses a state circle running downstream. Because its other endpoint is on the opposite side of that state circle, it must cross the state circle again. But this contradicts Lemma 3.11 (Utility lemma), as it first crossed running downstream.

Next suppose that σ_1 (or σ_2) crosses a non-prime arc. Again since its other endpoint is on the opposite side of the half-disk bounded by the non-prime arc and the segment of state circle between its endpoints, σ_1 (or σ_2) must cross back out of this non-prime half-disk. Since σ_1 (σ_2) is assumed to be simple, it may only exit the region by running downstream across the state circle. As in the previous paragraph, this leads to a contradiction to the Utility lemma.

Thus the arcs σ_1 and σ_2 must run adjacent to the segment of the 2-edge loop, as desired. This finishes the proof Case 1.

Case 2: S' is inessential in the lower polyhedron. By choice of orientation on our vertices, the only way S' can be inessential is if both of its white arcs cut off the same vertex with opposite orientation. Thus one of the arcs, say β_V, cuts off vertices to both sides, and thus lies in a bigon face. Hence there is a 2-edge loop in the upper polyhedron, and tracing back through the clockwise map as above, we conclude that the boundary of the EPD encircles the bigon, and the 2-edge loop is of type $\mathscr{B}$ again.   □

6.3   Step 2: Analysis Near Vertices

In this section, we will complete Step 2 of the outline given earlier. The main result here is Proposition 6.10, which shows that either Theorem 6.4 holds, or the polyhedral region near the arcs β_W and β_V have a very particular form. Before we can state this result, we need two auxiliary lemmas concerning directed arcs in shaded faces.

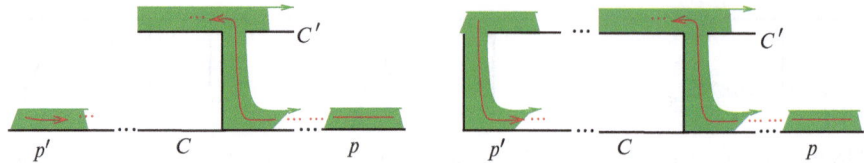

Fig. 6.4 *Left*: Hypothesis of Lemma 6.8. *Right*: Conclusion of the lemma: There is a 2-edge loop and σ runs adjacent to both segments of the loop

Lemma 6.8 (Adjacent loop). *Let σ be a directed simple arc contained in a single shaded face, adjacent to a state circle C at a point p on C. Suppose σ runs upstream across a state circle C' after leaving p, but then eventually continues on to be adjacent to C again at a new point p'. Then σ must run adjacent to two distinct segments of H_A connecting C to C'.*

Lemma 6.8 is illustrated in Fig. 6.4.

Proof. Since σ runs from C, through C', and eventually back to C, it must cross C' twice. Lemma 3.11 (Utility lemma) implies it first crosses C' running upstream, adjacent to some segment connecting C and C', then runs downstream. When it runs downstream, it must run adjacent to a segment connecting C' to some state circle C''. We show C'' must be C. Then, since σ is simple, the two segments connecting C to C' must be distinct, and we have the result.

Suppose C'' is not C. Because σ must run adjacent to C further down the directed arc, σ must leave C''. Recall that the only possibilities are that σ runs over a non-prime switch or runs downstream across C''. If downstream across C'', then it must cross C'' again. Lemma 3.11 (Utility lemma) implies this is impossible, as it is running downstream for the first crossing. If σ runs over a non-prime switch without crossing into the half-disk bounded by the non-prime arc, then on the opposite side it is adjacent to C'' again, and we have no change. If it crosses into the non-prime half-disk bounded by C'' and the non-prime arc and exits out again, then Lemma 3.7 (Shortcut) implies that it exits running downstream across C'', which again gives a contradiction.

So the only remaining possibility is that σ crosses into the non-prime half-disk and does not exit out again. This means C must be contained in this half-disk. But C' is on the opposite side, since σ leaves the region containing C' when crossing the non-prime arc. This is impossible: A segment connects C to C', hence C and C' must be on the same side of the non-prime arc. So C'' must equal C. $\qquad\square$

Lemma 6.9. *Suppose there are arcs σ_1 and σ_2 in distinct shaded faces in the upper polyhedron, but that each runs adjacent to points (in a neighborhood of points) p_1 and p_2 on the same state circle C. Then either*

(1) *At least one of σ_1 or σ_2 runs upstream across some other state circle and Lemma 6.8 applies; or*

(2) *Both arcs remain adjacent to the same portion of C between p_1 and p_2.*

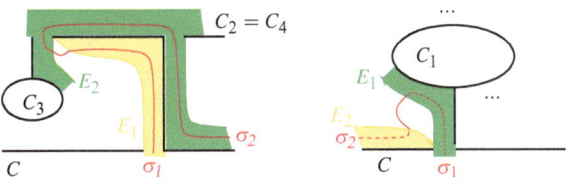

Fig. 6.5 Conclusion of step 2 in the proof of Theorem 6.4: Either there is a 2-edge loop, or each polyhedral region containing β_V, β_W is one of these two forms, with the specified colors

In other words, Lemma 6.9 says that if neither σ_1 nor σ_2 cross a state circle between p_1 and p_2, then they cannot run over non-prime switches, either. They can, in fact, intersect a single endpoint of a non-prime arc. But they cannot run adjacent to both endpoints of a non-prime arc.

Proof. If one of σ_1, σ_2 crosses another state circle between points p_1 and p_2 there is nothing to prove. Suppose that neither σ_1 nor σ_2 cross another state circle between points p_1 and p_2; then each is embedded in the complement of the graph H_A. Form a simple closed curve meeting H_A exactly twice by connecting the portions of σ_1 and σ_2 between p_1 and p_2 with small arcs crossing C at p_1 and p_2. Replacing segments of H_A with crossings, this gives a simple closed curve in the diagram of the link meeting the link transversely exactly twice. Because the diagram is assumed to be prime, the curve must contain no crossings on one of its sides. Since each side contains a portion of C between p_1 and p_2 in H_A, one of those portions must not be connected to any segments of H_A. Because non-prime arcs are required to bound segments on both sides, this means there are no non-prime arcs attached to this portion of C as well. Then a single tentacle runs adjacent to each side of this portion of C, and because shaded faces are simply connected, σ_1 must run through one, and σ_2 must run through the other. □

We are now ready to state and prove the main result of this section.

Proposition 6.10. *With the hypotheses of Theorem 6.4, either*

(1) *The conclusion of the theorem is true and we have a 2-edge loop of type $\mathscr{A}$, $\mathscr{B}$, $\mathscr{C}$, or $\mathscr{D}$, or*
(2) *The polyhedral regions containing β_V and β_W are of one of two forms, shown in Fig. 6.5.*

In both cases in the figure, σ_2 immediately leaves the polyhedral region, either through a non-prime arc, or by crossing some state circle.

Proof. By Lemma 6.7, we may assume that β_V and β_W, in white faces V and W respectively, are in distinct polyhedral regions. Then Lemma 4.15 (Entering polyhedral region) implies that if we direct σ_1 toward β_W, it first enters the region containing W running downstream across a state circle, which we denote C_W, while σ_2 enters the region of W either running upstream across C_W, or across a non-prime

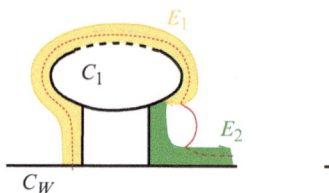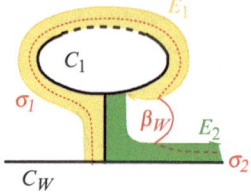

Fig. 6.6 Case: Head of E_2 meets tail of E_1, σ_2 runs directly to β_W. On *left*: obvious 2-edge loop. *Right*: nugatory crossing contradicts prime. The *dashed line* on the state circle indicates that there may be non-prime switches

arc. In either case, σ_1 connects immediately to β_W, that is, without crossing any additional state circles or non-prime arcs.

Let E_1 be the ideal edge of the polyhedral decomposition on which σ_1 meets β_W. Note E_1 is a directed edge, with its head on C_W and tail on some other state circle connected to C_W by a segment. Let E_2 denote the ideal edge on which σ_2 meets β_W. Since β_W cuts off a single ideal vertex, either the head of E_2 meets the tail of E_1, or vice versa. We must consider both cases.

Case 1: Suppose σ_2 connects immediately to β_W upon entering the polyhedral region of W, that is, without crossing any additional state circles. As noted above, the ideal edge E_1 has its head on C_W, runs adjacent to a segment which we denote s_1 connecting C_W to a state circle C_1, then has its tail on C_1. The arc σ_1 runs across C_W and adjacent to s_1. There are two subcases to consider.

Subcase 1a: The head of E_2 meets the tail of E_1. Then the head of E_2 must also lie on the state circle C_1. Since σ_2 connects immediately to β_W by assumption, and since σ_2 is adjacent to C_W when it enters the region of W (by Lemma 4.15 (Entering polyhedral region)), E_2 has its tail on C_W and thus E_2 runs adjacent to a segment s_2 connecting C_1 and C_W. We may isotope β_W to cut off a very small portion of the white face W, forcing σ_2 to run adjacent to s_2. See Fig. 6.6, left.

Now, provided $s_1 \neq s_2$, we have found two segments connecting C_W and C_1 with the boundary of the EPD running adjacent to both, through distinct shaded faces on the segments. Note that in this case, the boundary of the EPD is of type $\mathscr{A}$ of the statement of the theorem. Thus option (1) in the statement of the proposition holds.

So suppose $s_1 = s_2$, so that we don't pick up this 2-edge loop. We will now show this leads to a contradiction to the fact that the diagram is prime. See Fig. 6.6, right. Form a loop in H_A by following σ_1 from the point where it is adjacent to the segment s_1 to β_W, then following β_W to σ_2, then following σ_2 to the point where it is adjacent to the segment s_2. Since $s_1 = s_2$, connect these into a loop by drawing a line through this segment connecting the endpoints. Call the loop γ. Because σ_1 connects immediately to β_W without running through additional state circles, γ is embedded in the complement of H_A, except where it crosses the segment $s_1 = s_2$. Replace all segments of H_A with crossings of the diagram, and push γ slightly off the crossing

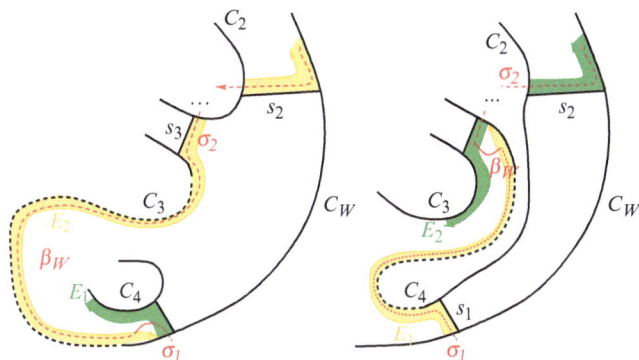

Fig. 6.7 Possibilities for R_W when met by an essential product disk

of the segment. The result is a loop meeting the diagram twice transversely with crossings on both sides, contradicting the fact that the diagram is prime.

Subcase 1b: The head of E_1 meets the tail of E_2. The head of E_1 lies on C_W. Hence the tail of E_2 must also lie on C_W. The general form of β_W and the ends of the arcs σ_1 and σ_2 where they connect to β_W is on the right of Fig. 6.5. Thus option (2) in the statement of the proposition holds for W.

Case 2: Suppose that σ_2 does not immediately connect to β_W. Lemma 4.16 implies that σ_2 crosses upstream into some state circle C_2, hence adjacent to some segment s_2 connecting C_W and C_2, then out of C_2 again running downstream, hence adjacent to some segment s_3 connecting C_2 and some state circle C_3, as in Fig. 4.5, on p. 65. At this point, σ_2 immediately meets β_W, without crossing any additional state circles or non-prime arcs. Hence the edge E_2 has its head on C_2 and its tail on C_3.

Subcase 2a: The head of E_1 meets the tail of E_2. Recall that the head of E_1 is on C_W, and the tail of E_2 is on C_3. In order for these edges to meet in this way, we must have $C_W = C_3$. But now, we have a segment s_2 connecting C_W to C_2, and σ_2 runs adjacent to this segment as it runs upstream into C_2. We also have a segment s_3 connecting C_2 to $C_3 = C_W$, and σ_2 runs adjacent to this segment as it runs downstream out of C_2 to meet β_W. In this case, s_2 cannot equal s_3, since σ_2 is assumed to be simple. So s_2 and s_3 form the two segments giving the desired 2-edge loop. This case is shown on the left of Fig. 6.7, where again the dashed line on the state circle $C_3 = C_W$ indicates that there may be non-prime switches. Note this is of type $\mathscr{B}$ in Theorem 6.4. Thus option (1) in the statement of the proposition holds for W.

Subcase 2b: The head of E_2 meets the tail of E_1. The tail of E_1 is on some state circle C_4 connected to C_W by a segment s_1, which σ_1 runs adjacent to. Since the head of E_2 is on C_2, C_2 must equal C_4. Then we have a segment s_1 connecting C to $C_4 = C_2$, with σ_1 adjacent to s_1, and a segment s_2 connecting C_W to $C_2 = C_4$, with σ_2 adjacent to s_2. Provided $s_1 \neq s_2$, this gives the desired result. This is shown on the right in Fig. 6.7. Note in this case our loop is of type $\mathscr{A}$ in the statement of Theorem 6.4.

Suppose $s_1 = s_2$. In this case, we will find a 2-edge loop of type $\mathscr{D}$ stacked on the opposite side of $C_2 = C_4$ from the arc β_W, or show that our diagram is as on the right of Fig. 6.5.

Now, we are assuming $s_1 = s_2$. Consider the circle $C_2 = C_4$. Both σ_1 and σ_2 are adjacent to this state circle at the point where the segment $s_1 = s_2$ meets it. Additionally, by shrinking β_W, we see that both σ_1 and σ_2 are adjacent again to it at the point where E_1 meets E_2. So Lemmas 6.9 and 6.8 imply either that σ_2 runs adjacent to distinct segments forming a 2-edge loop in $\mathbb{G}_A$—note such a loop will be of type $\mathscr{D}$, since the arc β_W is on the opposite side of a state circle meeting both segments of the loop—or there are no segments attached to $C_2 = C_4$ between these points of adjacency, and σ_1 and σ_2 run through tentacles adjacent to the state circle. In this latter case, we are on the left in Fig. 6.5.

In all cases we have shown that either (1) or (2) of the statement of the proposition is true for W. Since the argument is symmetric with respect to the two faces V and W, the proposition follows.   □

6.4   Step 3: Building Staircases

By Proposition 6.10, we may assume the polyhedral regions containing V and W each look like one of the diagrams of Fig. 6.5. We have two vertices, with corresponding arcs β_V and β_W, and two corresponding state circles C_V and C_W, respectively, with C playing the role of C_V, C_W, in Fig. 6.5. (Note that we may have $C_V = C_W$.) Consider first β_W. In Fig. 6.5, the arc σ_1 crosses C_W running downstream toward β_W. The arc σ_2, if it crosses C_W at all, must do so running upstream toward β_W.

Now direct σ_2 away from β_W. If it crosses C_W, it does so running downstream. We will use Lemma 3.10 (Downstream) to build a staircase of σ_2 away from β_W.

Lemma 6.11 (Building the first stair of a zig-zag). *Suppose $C_W \neq C_V$, and that σ_2, directed away from C_W, crosses C_W running downstream. Then either*

(1) *The conclusion of Theorem 6.4 holds and we have a 2-edge loop, or*
(2) *β_W is as on the left of Fig. 6.5, and σ_1 and σ_2 run parallel to the same two segments on either side of C_W, both of which are part of the same vertex, cut off by β_W.*

Proof. The arc σ_2 runs downstream across C_W, through a tentacle which is then adjacent to some state circle C. Say the tentacle has its head adjacent to a segment s_2 connecting C_W and C. Note that C might equal C_V.

Claim. The arc σ_1 must cross C running upstream, when running from β_W to β_V.

Proof of Claim: If C separates β_W and β_V, then σ_2 and σ_1 must cross C. Lemma 3.10 (Downstream) implies that σ_2 crosses in the downstream direction. Lemma 4.14 (Opposite sides) implies that σ_1 must cross C in the upstream direction, as claimed.

Now suppose C does not separate β_W and β_V. Because σ_2 is running downstream, if it crosses C it does so running downstream, and Lemma 3.11 (Utility) implies it cannot cross back, contradicting the fact that C does not separate. So in this case, σ_2 does not cross C.

Hence, either σ_2 terminates in the arc β_V, without crossing any non-prime arcs, or σ_2 must cross into a non-prime half-disk through a non-prime arc with endpoints on C, without exiting the half-disk. In the first case, the region of V is as on the right of Fig. 6.5, with σ_2 matching the labels in that figure, since $C_V \neq C_W$. Then notice σ_1 crosses C. Similarly, in the case that σ_2 enters a non-prime half-disk without exiting, β_V is inside the half-disk bounded by C and the non-prime arc, and so σ_1 must cross into this half-disk as well, and because the non-prime tentacle belongs to the shaded face of σ_2, σ_1 must cross through a tentacle running through C. In either case, σ_1 crosses C. Since C does not separate β_V and β_W, in fact σ_1 must cross C twice, first running upstream, then running downstream, by Lemma 3.11 (Utility). This finishes the proof of the claim.

To continue with the proof of the lemma, we change the direction of σ_1, so it is running across C in the downstream direction, when oriented from β_V to β_W. We may then apply Lemma 3.10 (Downstream) to σ_1, directed toward β_W, for note it will run downstream across C, and eventually downstream across C_W, exiting out of every non-prime half-disk along the way. Hence Lemma 3.9 (Staircase extension) implies that σ_1 defines a right-down staircase between C and C_W, with σ_1 running adjacent to each connecting staircase in the segment.

Arguing by A-adequacy of the diagram, similar to the proof of Lemma 3.14, the staircase of σ_1 consists of a single segment s_1, connecting C and C_W. Recall that σ_2 runs adjacent to a segment s_2, also connecting C and C_W. We will argue that either $s_1 \neq s_2$ and option (1) holds, or $s_1 = s_2$ and we are in option (2).

Case 1: Suppose that β_W is as on the right of Fig. 6.5. Direct σ_2 away from β_W. If σ_2 runs upstream across any other state circle before running downstream across C_W, then Lemma 6.8 will imply that there is a 2-edge loop of type $\mathscr{B}$. Hence we assume σ_2 does not run upstream across another state circle from the point where it leaves β_W to the point where it crosses C_W.

Similarly, consider σ_1 running (downstream) towards β_W from s_1 to cross C_W: If it runs upstream between s_1 and C_W, then Lemma 6.8 implies that there is a 2-edge loop of type $\mathscr{D}$.

So suppose σ_2 does not run upstream between β_W and crossing C_W, and suppose that σ_1 does not run upstream between leaving s_1 and crossing C_W. Then σ_1 and σ_2 are both adjacent to C_W at β_W. We claim that s_1 cannot equal s_2. In this case, s_1 and s_2 and the portions of the boundary of the EPD adjacent to them form a 2-edge loop of type $\mathscr{C}$ and we are in option (1) of the statement of the lemma. Suppose, on

Fig. 6.8 Lemma 6.12: either
we have desired 2-edge loop,
or the graph H_A is as shown

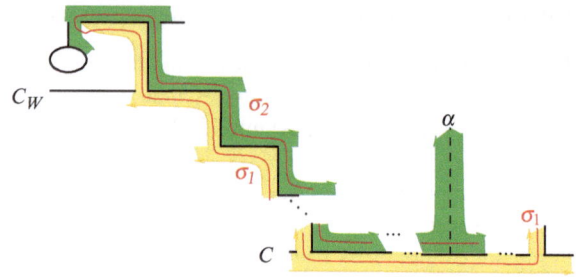

the contrary, that $s_1 = s_2$. Then s_1, s_2 are also adjacent to C_W where this segment
attaches to C_W. Lemma 6.9 implies that both σ_1 and σ_2 must run along C_W between
these two points. But note that σ_1 and σ_2 must run in opposite directions. Hence the
loop following σ_2 on one side of C_W, following σ_1 on the other side, connecting
where these are adjacent into a loop, becomes a loop in the diagram meeting the
diagram twice, bounding crossings on either side. This contradicts the fact that the
diagram is prime. Thus $s_1 \neq s_2$ as desired.

Case 2: Suppose that β_W is as on the left of Fig. 6.5. If σ_2, directed from W to
V, runs upstream before crossing C_W, then Lemma 6.8 implies there is a 2-edge
loop of type $\mathscr{F}$. If σ_1, directed from V to W, runs upstream between leaving s_1 and
crossing C_W, then there is a 2-edge loop of type $\mathscr{D}$. If $s_1 \neq s_2$, then there is a 2-edge
loop of type $\mathscr{A}$. If none of these three things happen, then $s_1 = s_2$, and as before,
Lemma 6.9 implies that σ_1 and σ_2 both run adjacent to C_W between the point where
$s_1 = s_2$ and the segment on the opposite side of C_W where the two arcs run adjacent
on opposite sides. In this case, the segment $s_1 = s_2$ is part of the same vertex as β_W,
as claimed in the statement of the lemma. $\square$

The previous lemma is the first step in creating a maximal right-down staircase
for the vertex corresponding to β_W. The next lemma gives the full staircase of a
zig-zag.

Lemma 6.12 (Full staircase). *Suppose $C_W \neq C_V$, and σ_1 and σ_2 are directed from
β_W to β_V, and that σ_2 crosses C_W running downstream. Then, either*

(1) *The conclusion of Theorem 6.4 holds; or*
(2) *β_W is as on the left of Fig. 6.5, and the vertex of β_W forms a right-down
    staircase, with σ_1 and σ_2 adjacent on either side.*

*In case (2) the staircase is maximal, in the sense that at the bottom of the right-down
staircase, σ_2 either crosses C_V, or crosses over a non-prime arc α with endpoints
on some state circle C, and does not exit the corresponding half-disk. In the latter
case, the arc σ_1 also crosses into this half-disk, first running upstream across C then
running downstream into the half-disk bounded by α and C. Additionally, σ_1 does
not cross any other state circles between its two crossings of C.*

The form of the graph H_A in case (2) of Lemma 6.12 is illustrated in Fig. 6.8.

Proof. By Lemma 6.11, we may assume that β_W is as on the left of Fig. 6.5, and σ_1 and σ_2 run parallel to the same segments on either side of β_W, which are both part of the vertex at β_W.

Claim. For $i = 1, 2$, σ_i defines a right-down staircase, with σ_i running adjacent to each segment of the staircase.

Proof of Claim: We may apply Lemma 3.10 (Downstream) to the arc σ_2, directed away from β_W. This lemma implies that σ_2 defines a right-down staircase, with σ_2 running adjacent to each segment of the staircase in a downstream direction, either until σ_2 crosses a non-prime arc without exiting the half-disk it bounds with a state circle C, or crosses C_V and runs to β_V.

If σ_2 crosses C_V, then C_V separates V and W, so σ_1 must also cross C_V, running upstream when directed to V. If σ_2 crosses a non-prime arc without exiting the half-disk it bounds with C, then note V must lie inside this half-disk, so σ_1 must cross into this half-disk. Moreover, C cannot separate V and W, so σ_1 actually must cross C twice, by Lemma 3.11 (Utility), first running upstream, then downstream when directed toward V.

In either case, σ_1 crosses the last state circle C of the staircase of σ_2 running upstream, directed toward V. Change the direction on σ_1. It runs downstream across C, and downstream across C_W, and must cross out of any non-prime half-disks between these. So Lemma 3.10 (Downstream) implies that σ_1 defines a right-down staircase, with σ_1 running adjacent to each segment of the staircase. This finishes the proof of the claim.

To continue with the proof of the lemma we note that adequacy implies that the segments of the staircases defined by σ_1 and by σ_2 must actually run between the same sequence of state circles.

Recall that we know that the first segment of the staircase, on the other side of C_W from β_W, is shared by both σ_1 and σ_2. Suppose we have shown that σ_1 and σ_2 run adjacent to the first k stairs of a right-down staircase forming the vertex of β_W, and that σ_2 runs to a $(k + 1)$-st step. We will show the theorem holds at this step.

The arc σ_2 runs from the top of the $(k + 1)$-st step, somewhere, then downstream adjacent to the segment of the step. If σ_2 runs upstream first, before running downstream, then Lemma 6.8 implies that a 2-edge loop of type $\mathscr{F}$ occurs. Similarly, when directed downstream, the arc σ_1 runs adjacent to the segment of the $(k + 1)$-st step, somewhere, and then downstream adjacent to the k-th step. If it runs upstream between the two segments, then Lemma 6.8 will imply there is a 2-edge loop, this time of type $\mathscr{G}$. In both cases option (1) holds.

Now assume neither σ_1 nor σ_2 runs upstream between the segments of these steps. If the segments of σ_1 and σ_2 at this $(k + 1)$-st step are distinct, then we have a 2-edge loop of type $\mathscr{A}$; again option (1) holds.

If the segments are not distinct, then Lemma 6.9 implies that σ_1 and σ_2 both run adjacent to the state circle of this step between the two segments of the step. Thus this step is a continuation of the same vertex corresponding to β_W.

By induction, either (1) holds or σ_1, σ_2 share every segment of the staircase.

Finally, suppose the staircase ends with σ_2 entering into the non-prime half-disk bounded by a non-prime arc and the state circle C, without exiting. We have already seen that σ_1 must also cross C in this case, first upstream and then downstream. Suppose σ_1 crosses additional state circles between these two crossings of C. Then Lemma 6.8 implies that there is a 2-edge loop of type $\mathscr{B}$ on the underside of the state circle C.   □

Lemma 6.13. *Suppose σ_2 runs across C_W in the downstream direction, out of every non-prime half-disk that it enters, and terminates with σ_2 crossing C_V. Then the conclusion of Theorem 6.4 holds.*

Proof. By Proposition 6.10 we reduce to the case that β_W and β_V are as in Fig. 6.5. The colors on these figures are fixed, given our choice of direction in which to pull β_W and β_V off their corresponding vertices (Lemma 6.1, EPD to oriented square).

This means that both vertices cannot be of the same form in that figure, or a green[2] shaded face would lie adjacent to both sides of the same state circle, which is impossible by Lemma 3.4 (Escher stairs).

Thus one vertex, β_W say, is as on the left of Fig. 6.5, and the other vertex, β_V, is as on the right. Relabel so σ_1 runs through the orange face and σ_2 runs through the green.

By Lemma 6.12, we reduce to the case that either $C_W = C_V$, or both σ_1 and σ_2 run adjacent to either side of a maximal right-down staircase from C_W to C_V, corresponding to the vertex of β_W. In both cases, when σ_1 and σ_2 are directed toward C_V, they run adjacent to the same segment which meets C_V on the opposite side of that containing β_V: if $C_W = C_V$, then σ_1 and σ_2 are adjacent to the segment shown on the left in Fig. 6.5; otherwise they are adjacent to the last segment of the staircase from C_W to C_V.

After leaving this segment, σ_1 and σ_2 split up and run to β_V. Thus the two are adjacent to each other and to C_V in two distinct points: at a segment on one side of C_V, and at β_V on the other side.

If σ_1 crosses C_V (running upstream) then runs upstream again before meeting β_V, Lemma 6.8 implies there will be a 2-edge loop of type $\mathscr{B}$. Similarly, if σ_2 runs upstream before crossing C_V, then there will be a 2-edge loop, and this must be of type $\mathscr{F}$, as σ_2 is running downstream to C_V, so must pass through a non-prime switch to run upstream.

If neither σ_1 nor σ_2 run upstream between the point where they leave the segment of the zig-zag of β_W and the point where they meet again at β_V, then a simple closed curve crosses the knot diagram at the base of the zig-zag and at β_V, and nowhere else, following σ_1 on one side and σ_2 on the other, and encircles crossings on both sides. This contradicts the hypothesis that the diagram is prime.   □

---

[2]In grayscale versions of this monograph, green will appear darker gray, orange lighter gray.

6.5 Step 4: Inside Non-prime Arcs

At this point in the proof of Theorem 6.4, we either have the conclusion of the theorem, or we have specialized to cases where the form of the graph H_A is very restricted. In particular,

- β_V and β_W must be in distinct polyhedral regions (Step 1);
- The graph near β_W and β_V must be of one of the two forms shown in Fig. 6.5 (Step 2);
- σ_2 runs down a (possibly empty) maximal right-down staircase and across a non-prime arc, as in Fig. 6.8 (Step 3).

To finish the proof, we need to analyze what happens to the EPD when β_V and β_W are separated by a non-prime arc α.

Lemma 6.14. *Suppose β_V and β_W are separated by a non-prime arc α, with the arc σ_2, say, crossing α. Suppose α is outermost among all such arcs, with respect to β_W. That is, α is the first such non-prime arc crossed by σ_2 when directed toward β_V. Then we have the conclusion of Theorem 6.4.*

Proof. We break the proof into two cases: first, that σ_2 does not run upstream after crossing α, and second, that it does run upstream.

Case 1: Suppose σ_2 does not run upstream after crossing α. Now suppose, by way of contradiction, that the conclusion of Theorem 6.4 is not true. We will find a contradiction to primeness of the diagram.

Since σ_2 does not run upstream after crossing α, it will not run downstream either, for to run downstream would be to cross the state circle C out of the non-prime half-disk bounded by α, contradicting the hypotheses. Therefore, after crossing α, σ_2 must run directly to β_V without crossing any additional state circles. We know the graph H_A must have one of the forms of Fig. 6.5, and that σ_2 cannot cross an additional state circle after entering the region of β_V, hence β_V must be as on the right of that figure, so α is an arc in an orange face.

Next, Lemma 6.12 (full staircase) implies that on the opposite side of α, in the region containing β_W, σ_1 and σ_2 run adjacent to the same (possibly empty) right-down staircase corresponding to the vertex of β_W. However, notice that if the staircase is non-empty, then β_W must have the form of the left of Fig. 6.5, and the colors must be as in Fig. 6.8. That is, α is an arc in a green face. But in the previous paragraph, we argued that α is in an orange face. This is a contradiction. So the zig-zag of β_W must be empty, and β_W must have the form of the right of Fig. 6.5. Note that this implies that σ_2 meets no state circles on either side of the non-prime arc α. See Fig. 6.9.

By Lemma 6.12, σ_1 crosses C twice, but meets no state circles between these crossings. Then the boundary of the EPD gives a simple closed curve in the diagram which meets the diagram exactly twice, once each time σ_1 crosses C. This contradicts the fact that the diagram is prime.

Fig. 6.9 The contradiction in Lemma 6.14, Case 1: β_W and β_V both have the form of the right of Fig. 6.5, and σ_2 meets no state circles

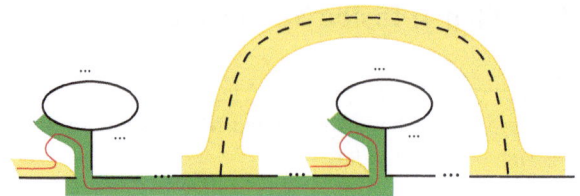

Case 2: The arc σ_2 does run upstream after crossing α, say across some state circle C_1.

If σ_2 runs back to C from C_1, then Lemma 6.8 (Adjacent loop) implies there is a 2-edge loop of type $\mathcal{F}$.

If not, then we claim that the theorem holds or σ_1 must also run adjacent to a segment connecting C and C_1. This can be seen as follows. First, if C_1 separates β_V and β_W, then σ_1 must also cross C_1. Since σ_1 is running downstream (by Lemma 4.14 (Opposite sides)), Lemma 3.10 (Downstream) implies that it must run adjacent to a segment from C to C_1, as desired. If C_1 does not separate, then σ_2 must cross it twice, the second time running downstream to some C''. If $C'' = C$, then we must have a 2-edge loop of type $\mathcal{F}$. If $C'' \neq C$, consider σ_1. It runs downstream across C, along a segment connecting C to some C'. If $C' \neq C_1$, then $C' = C''$, else we could not connect ends of σ_1 and σ_2 at β_V. But now, σ_1 and σ_2 cannot cross $C' = C''$, or we would build two staircases from C' contradicting Lemma 3.14 (Parallel stairs). On the other hand, β_V cannot lie on $C' = C''$, since it would lie at the tails of two tentacles, which do not meet at a vertex. The only possibility is $C_1 = C'$, as desired.

Thus σ_1 and σ_2 both run adjacent to segments from C to C_1 inside α. If these segments are distinct, we have a 2-edge loop of type $\mathcal{E}$.

If not, we will show we have a contradiction to the fact that the diagram is prime. By assumption, σ_1 and σ_2 are adjacent to the vertex corresponding to β_W just outside α on C, and they meet no additional state circles outside α. If they are also adjacent to the same segment inside α, then we may form a loop in the diagram meeting C twice, meeting no other state circles, by following σ_1 on one side and σ_2 on the other. This will descend to a loop in the diagram enclosing curves on both sides, meeting the diagram just twice, contradicting the fact that the diagram is prime. $\square$

Completion of the proof of Theorem 6.4. As discussed in the beginning of this section, Steps 1 through 3 imply the theorem in all cases where β_W, β_V are not separated by a non-prime arc. Recall that by Step 3, either σ_2 runs directly from β_W across a non-prime arc separating V and W, or we have σ_1 and σ_2 adjacent to a maximal right-down staircase of the vertex of β_W, as in Fig. 6.8. By Lemma 6.12, in the latter case σ_2 must also cross a non-prime arc separating the β_W and β_V. Hence, in all cases, there is a non-prime arc that separates β_W, β_V. Now we pass to an outermost such non-prime arc, and apply Lemma 6.14 to obtain the conclusion. $\square$

Chapter 7
Diagrams Without Non-prime Arcs

In this chapter, which is independent from the remaining chapters, we will restrict ourselves to A-adequate diagrams $D(K)$ for which the polyhedral decomposition includes no non-prime arcs or switches. In this case, one can simplify the statement of Theorem 5.14 and give an easier combinatorial estimate for the guts of M_A. This is done in Theorem 7.2, whose proof takes up the bulk of the chapter.

Definition 7.1. In the A-adequate diagram $D(K)$, let b_A denote the number of bigons in the A-regions of the diagram. (Recall Definition 5.16 on p. 86 for the notion of an A-region.) In other words, in Fig. 5.4, b_A is the number of bigons in twist regions where the A-resolution is short. Define

$$m_A = e_A - (e'_A + b_A).$$

Since each bigon of b_A corresponds to a redundant edge of the graph G_A, the quantity m_A is always non-negative.

Note that the quantity m_A counts the number of distinct segments of H_A that connect the same state circles, excepting those segments that come from twist regions and bound simple rectangles in H_A. In other words, $m_A = 0$ precisely when every 2-edge loop in $\mathbb{G}_A$ has edges belonging to the same twist region, as in Corollary 5.19 on p. 88.

The main result of this chapter extends the simple diagrammatic statement of Corollary 5.19 to a context where the corollary does not directly apply.

Theorem 7.2. *Let $D(K)$ be a prime, A-adequate diagram, and let S_A be the essential spanning surface determined by this diagram. Suppose that the polyhedral decomposition of $M_A = S^3 \backslash\backslash S_A$ includes no non-prime arcs; that is, no further cutting was required in Sect. 2.3. Then*

$$\chi_-(\mathbb{G}'_A) - 8m_A \leq \chi_-(\mathrm{guts}(M_A)) \leq \chi_-(\mathbb{G}'_A),$$

where the lower bound is an equality if and only if $m_A = 0$.

D. Futer et al., *Guts of Surfaces and the Colored Jones Polynomial*, Lecture Notes in Mathematics 2069, DOI 10.1007/978-3-642-33302-6_7,
© Springer-Verlag Berlin Heidelberg 2013

To derive Theorem 7.2 from Theorem 5.14 on p. 84, it suffices to bound the number $\|E_c\|$ of complex disks required to span the I-bundle of the upper polyhedron. (See Definition 5.9 on p. 81.) Note that by Theorem 6.4 on p. 93, each disk $D \in E_c$ must run along a 2-edge loop of $\mathbb{G}_A$. If this loop corresponds to a single twist region, as in Corollary 5.19 on p. 88, then a disk corresponding to this loop cannot be complex. In the following argument, we will bound $\|E_c\|$ in terms of m_A, where m_A accounts for the loops that *do not* correspond to twist regions.

Before diving into the proof of Theorem 7.2, we give a sample application.

Example 7.3. Lickorish and Thistlethwaite introduced the notion of a *strongly alternating tangle* [61]. This is an alternating tangle T, such that both its numerator and denominator closures are alternating, prime, reduced diagrams. (See Definition 8.1 on p. 120 for the notions of numerator, denominator, and tangle sum.) A *semi-alternating diagram* D is the numerator closure of the tangle sum $T_1 + T_2$, where each T_i is strongly alternating but their sum $T_1 + T_2$ is non-alternating. Lickorish and Thistlethwaite observed that these diagrams are both A- and B-adequate.

If D is a twist-reduced, strongly alternating diagram, there is exactly one state circle C of $s_A(D)$ that runs through both tangles T_1 and T_2. In the all-A resolution of T_1 (resp. T_2), this state circle appears as a pair of arcs along the north and south (resp. east and west) of the tangle. Then, a 2-edge loop in $\mathbb{G}_A(D)$ can take one of two forms. The two edges of this loop either belong to a single twist region (in which case they do not contribute to m_A), or else they form a *bridge* of two edges that spans the tangle north to south, or east to west. (See Fig. 8.6 on p. 129 for an example.) The quantity m_A is then exactly equal to the number of bridges in the tangles. Thus, applied to a semi-alternating diagram, Theorem 7.2 has the simpler formulation

$$\chi_-(\mathbb{G}'_A) - 8 \text{ (number of bridges in } \mathbb{G}_A) \leq \chi_-(\text{guts}(M_A)) \leq \chi_-(\mathbb{G}'_A).$$

7.1 Mapping EPDs to 2-Edge Loops

Recall that Theorem 6.4 shows that every EPD in the upper polyhedron determines a normal square of one of seven types ($\mathscr{A}$ through $\mathscr{G}$), as shown in Fig. 6.1. Under the hypothesis that the polyhedral decomposition includes no non-prime arcs or switches, we will simplify these seven cases to three (see Fig. 7.1).

Definition 7.4. A *brick* is a pair of segments s, s' of the graph H_A that connect the same state circles C_0 and C_1.

Note that a closed curve in the projection plane consisting of segments s and s', as well as parallel arcs of C_0 and C_1 from s to s', is topologically a rectangle. This is the origin of the term *brick*. We will always depict bricks with state circles horizontal and segments of H_A vertical, as in Fig. 7.1.

Note as well that the segments s and s' split the annular region between C_0 and C_1 into two rectangular components. We will say that tentacles adjacent to s and s' are on the *same side* of the brick if these tentacles lie in the same rectangular

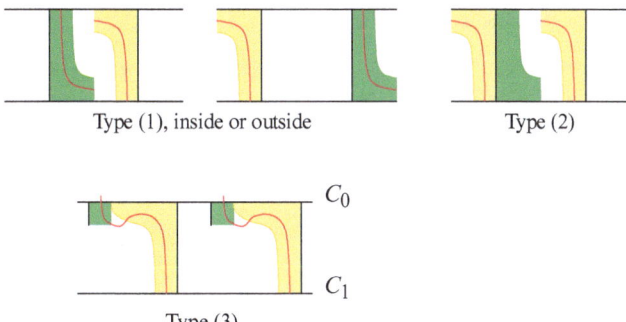

Type (1), inside or outside   Type (2)

C_0

C_1

Type (3)

Fig. 7.1 When there are no nonprime arcs, each EPD can be associated with a 2-edge loop of type (1), (2), or (3), illustrated

component, and that these tentacles are on *opposite sides* of the brick if they belong to different components.

Definition 7.5. Any EPD meets exactly two distinct shaded faces. We assign each shaded face a unique color. A *color pair* of an EPD is a choice of two distinct shaded faces met by a single EPD.

By Lemma 5.8 on p. 80, any tentacle caries at most two EPDs with the same color pair.

Proposition 7.6. *Let $D(K)$ be a prime, A-adequate diagram of a link in S^3 with prime polyhedral decomposition of $S^3 \backslash\backslash S_A$ such that the polyhedral decomposition contains no non-prime arcs. Let E be an EPD embedded in the upper polyhedron, with associated normal square of Lemma 6.1 (EPD to oriented square), and denote the color pair of E by orange–green,[1] with the orientation convention of Lemma 6.1 (EPD to oriented square). Then we may associate E with a brick in H_A of one of the following forms.*

(1) *The normal square of E runs through distinctly colored tentacles adjacent to the segments of the brick (hence one orange, one green tentacle), with these tentacles lying on the same side of the brick.*

(2) *The normal square of E runs through two orange colored tentacles adjacent to the segments of the brick, necessarily on opposite sides of the brick, and in addition, a green tentacle is adjacent to one of the two segments.*

(3) *The normal square of E runs through two orange colored tentacles adjacent to the segments of the brick, necessarily on opposite sides of the brick. Moreover, the arcs of ∂E in these orange tentacles run to the tail of the orange tentacles on the state circle C_0 of the brick, there meet a vertex, and then run into green tentacles and across the state circle C_0.*

---

[1] Note: For grayscale versions of this monograph, orange faces in the figures will appear light gray, and green ones will appear darker gray.

The three possibilities are illustrated in Fig. 7.1.

Proof. This follows from an analysis of the normal squares of types $\mathscr{A}$ through $\mathscr{G}$ in the conclusion of Theorem 6.4. Notice that the normal squares in types $\mathscr{E}$, $\mathscr{F}$, and $\mathscr{G}$ include non-prime arcs as essential portions of the diagram, so none of these can occur in the setting at hand.

Consider first the normal squares of types $\mathscr{A}$ and $\mathscr{C}$, illustrated in Fig. 6.1 on p. 94. Note that the boundary of the EPD in these cases runs in tentacles of distinct colors adjacent to the 2-edge loop. Moreover, note that when we close off the 2-edge loop to form a brick, these two distinguished tentacles are on the same side of the brick. Hence we have type (1) in these cases.

Next consider type $\mathscr{B}$. The zig-zag at the top left of the figure showing type $\mathscr{B}$ in Fig. 6.1 is schematic, to represent the fact that there may be 0 or more segments in that zig-zag. If there are 0 segments in the zig-zag, the arc of the EPD in the green face may run either upstream or downstream from the top left. To prove this proposition, when we have a 2-edge loop of type $\mathscr{B}$, we need to condition on whether the arc of the EPD in the green, top left, runs upstream or downstream from this point.

Suppose first that it runs downstream. Then by Lemma 3.10 (Downstream), it must run downstream until it terminates. Notice that by our orientation convention (Lemma 6.1, EPD to oriented square), the arc cannot terminate in the tail of a green tentacle. Hence the arc σ_1 in the green must cross the top state circle C_0 in the brick of the 2-edge loop before it terminates. However, notice σ_1 cannot cross C_0 to the left of the left-most segment of our 2-edge loop, else it would force the orange segment to terminate, cutting off the arc in the orange. Thus the green terminates to the right of that left-most segment. This means that the tentacle adjacent to the right of that left-most segment must be green, and our brick is of type (2) in the statement of the proposition.

Next suppose that we have a 2-edge loop of type $\mathscr{B}$, but our arc in the top left in the green runs upstream rather than down, adjacent to a segment s_1. Then we have zero segments in the zig-zag vertex at the top left of type $\mathscr{B}$, and the state circle at the top of the brick, call it C_0, is connected by s_1 to some other state circle C'. Now, consider the arc σ_2 of the EPD in the orange tentacle on the right, oriented so that it is running downstream toward C_0. Either σ_2 must run downstream across C_0, adjacent to a segment s_2 connecting C_0 and C', or the arc σ_1 in the green tentacle must eventually run downstream to meet C_0. In the first case, we obtain a brick between C_0 and C', with the EPD running over distinctly colored tentacles on the same side of the brick, and we have a brick of type (1) in the statement of the proposition. In the second case, the arc σ_2 terminates at a vertex on C_0, and we have a brick of type (3). This finishes the proof in the case that our 2-edge loop coming from Theorem 6.4 is of type $\mathscr{B}$.

It remains to show that the proposition holds when our 2-edge loop is of type $\mathscr{D}$, shown in Fig. 6.1 on p. 94. The argument in this case requires three steps, illustrated in Fig. 7.2.

Fig. 7.2 Tentacles of an EPD of type $\mathscr{D}$, in the absence of non-prime arcs

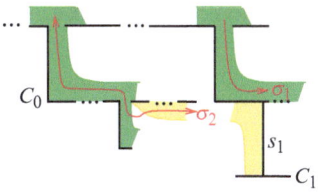

Step 1: Consider the arc σ_2 that lies in an orange tentacle at the bottom of Fig. 7.2. We claim that σ_2 runs upstream. To prove this claim, we need to show that σ_2 cannot run downstream, or terminate.

Suppose σ_2 runs downstream, across the state circle C_0 in the bottom of Fig. 7.2. Then, observe that the arc σ_1 in the green tentacle on the right of the figure is running downstream, on the opposite sides of C_0. Since σ_1 can only continue downstream until it terminates (by Lemma 3.10 (Downstream)), it must terminate immediately and connect to σ_2 at an ideal vertex. But such a vertex would be oriented green–orange–white (counter-clockwise), contradicting the orientation convention of Lemma 6.1 (EPD to oriented square).

Next, suppose that σ_2 terminates immediately, rather than running upstream. In that case, the arc σ_1 must run downstream across C_0 and immediately connect to meet the tail of σ_2 at a vertex. But this is impossible: the orange tentacle has only one tail, and this tail already forms a portion of the other vertex of the EPD, as illustrated in the figure. This proves the claim: σ_2 must run upstream, adjacent to some segment s_1 connecting C_0 to a state circle C_1.

Step 2: Now, consider the arc σ_1 lying in the green tentacle on the right of Fig. 7.2. This arc must run downstream across C_0 by orientation reasons, but it might either terminate immediately on the opposite side of C_0, or continue adjacent to a segment running to C_1.

If σ_1 terminates, then it meets an orange arc just on the opposite side of C_1. Lemma 6.8 (Adjacent loop) implies that σ_2 runs adjacent to some segment s_2 connecting C_0 to C_1, and s_1 and s_2 form a 2-edge loop of type (3) of the proposition.

Next, suppose that the arc σ_1 runs downstream. Then Lemma 3.11 (Utility) implies that it runs adjacent to a segment s_2 connecting C_0 to C_1. If s_1 and s_2 are distinct segments, then they form a 2-edge loop of type (1) as in the statement of the proposition, with arcs of the normal square running on the same side of the brick, in tentacles of distinct color. This proves the proposition in the case where s_1 and s_2 are distinct segments.

Step 3: Suppose that s_1 and s_2 are the same segment. Then we will repeat the argument as above and eventually end up with a brick of type (1) or (2), using induction and finiteness of the graph H_A, as well as primeness.

First, we claim the arc in the orange must again run upstream, for if it runs downstream we pick up a vertex of the wrong orientation, and if it terminates, then we get a contradiction to primeness: the arc in the orange connects across the bottom state circle to the arc in the green, and they form a loop meeting that state circle just once more, which gives a loop meeting the diagram twice with crossings on each side. Hence the orange arc runs upstream, say adjacent to a segment s.

The green arc either runs downstream, or terminates. If it terminates, it meets the tail of another orange tentacle, and Lemma 6.8 (Adjacent loop) implies that the orange runs down this tentacle, adjacent to some other segment s'. The segments s and s' form a 2-edge loop. Notice in this case that the green must terminate to the right of the segment s, else would cut off the orange tentacle. Thus the segment s must have a green tentacle adjacent to it on its right, and we have type (2).

Suppose the green arc runs downstream rather than terminating. Then it does so by running adjacent to a segment s'. Again s and s' form the desired brick of the proposition, of type (1) if they are distinct. If not, repeat verbatim the argument above, starting from the beginning of Step 3. By induction, we eventually get the brick given by the proposition.   □

7.2 A Four-to-One Mapping

For each essential product disk E in the upper polyhedron, Proposition 7.6 gives a mapping from E to some brick of H_A, with the E running through tentacles as in type (1), (2), or (3). Thus, when the EPDs are selected from the spanning set E_c of Lemma 5.8 on p. 80, Proposition 7.6 gives a function

$$f : E_c \to \{\text{bricks of type (1), (2), or (3)}\}.$$

The goal of this section is to show that the function f is at most four-to-one.

Definition 7.7. We say that a brick (a pair of segments between the same state circles of H_A) *supports* an essential product disk $E \in E_c$ if the function f above maps E to the given brick.

Lemma 7.8. *A single brick of H_A cannot support both an EPD of type (1) in one color pair and an EPD of type (3) in a different color pair.*

Proof. Let s, s' be the segments of the brick. Let E_3 be the EPD of type (3). For ease of exposition, we will assume that the shaded faces of E_3 are colored green and orange, with the orientation given by Lemma 6.1 (EPD to oriented square). Hence, the tentacles of E_3 look identical to those in Fig. 7.1, type (3).

Let E_1 be the EPD of type (1), also supported by the brick of s and s'. Note that since each of s and s' is adjacent to an orange tentacle, one of the shaded faces through which E_1 runs must be orange. Say that the color pair of E_1 is orange–blue, with the blue tentacle adjacent to segment s'.

Now, the orange tentacles adjacent to s, s' terminate with their tails on the heads of green tentacles on the state circle C_0. Since E_1 does not meet a green face, the arc σ_1 of E_1 running through the orange tentacle adjacent to s must run downstream

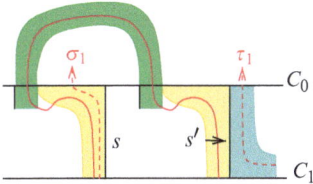

Fig. 7.3 When a single brick supports both a disk E_1 of type (1) and a disk E_3 of type (3), their intersections with a state circle C_0 must interleave. This will imply that E_1 must cut across the green (*darker shaded*) face containing E_3, which is a contradiction

across C_0 at a segment attached to C_0, to the right of the point where the orange tentacle terminates in a tail. Since the EPD E_3 runs through the tail of this tentacle, the arc of E_1 running through the same tentacle as the arc of E_3 must cross C_0 to the right of the point where that arc of E_3 crosses it.

On the other hand, the blue tentacle that E_1 runs through, adjacent to a segment s', must be on the right of s'. Hence the arc τ_1 of E_1 running through the blue tentacle crosses C_0 to the right of the arc of E_3 running adjacent to that same segment s'. See Fig. 7.3.

We conclude that E_1 and E_3 intersect state circle C_0 at interleaving points. We will see that this interleaving implies that the orange–blue disk E_1 must intersect the green shaded face, which will give a contradiction.

Let ρ be the arc of E_3 in the green face. We know, from Fig. 7.1 (3), that ρ crosses C_0 at two places, adjacent to segments s and s'. If we orient ρ from s to s', then it crosses C_0 first going upstream, then going downstream. By the Utility Lemma 3.11, these are the only intersections of ρ with C_0. Thus the green face separates the two tentacles of E_1 adjacent to segments s and s'.

Now, consider the intersections between state circle C_0 and arcs σ_1, τ_1 of E_1. As we have seen, the arc σ_1 of E_1 crosses C_0 running downstream. By Lemma 3.10 (Downstream), σ_1 keeps running downstream until it meets a white face W. Now, orient τ_1 toward face W, and consider the last time that it crosses C_0 before reaching face W. By Lemma 3.14 (Different streams), τ_1 must cross C_0 running upstream. But we already know a place where τ_1 crosses C_0, namely in the tentacle to the right of segment s'. By the Utility Lemma 3.11, τ_1 cannot cross C_0 twice running upstream. Thus τ_1 must cross C_0 to the right of s' and continue to white face W. This is a contradiction, since W lies on the other side of the green face from s'. $\square$

Lemma 7.9. *Suppose that a brick formed by segments s and s' supports an EPD E_2 of type (2), as well as an EPD E_3 of type (3) in a different color pair than that of E_2. Then E_2 and E_3 must run through all four tentacles adjacent to the brick. In particular, this will happen only if the two segments are adjacent to tentacles of the same two colors.*

Proof. As in the proof of Lemma 7.8, we will assume that the color pair of E_3 is green–orange, and that the brick of E_3 is positioned exactly like the brick of Fig. 7.1 (3), with identical colors.

Suppose for a contradiction that E_2 also runs through orange tentacles in that brick, but the color pair of E_2 is blue–orange. Consider the state circle C_0. Each orange tentacle of the brick terminates with its tail at a green tentacle on C_0. Because E_2 does not meet the green face, each of the arcs of ∂E_2 in the orange tentacles must run downstream across C_0, to the right of the point where the orange tentacle terminates. Thus, as in the proof of Lemma 7.8, we conclude that the intersection points of $\partial E_2 \cap C_0$ must interleave with the points of $\partial E_3 \cap C_0$.

Let σ be the arc of E_3 in the orange face. The interleaving intersections with C_0 mean that two points of $E_1 \cap C_0$ lie on opposite sides of σ. Thus the arc of E_1 in the orange face must intersect $\sigma \subset E_3$. By Lemma 4.9 on p. 60, it follows that (the normal square of) E_1 must also intersect (the normal square of) E_3 in another shaded face. This contradicts the hypothesis that E_1 is orange–blue while E_3 is orange–green. □

Lemma 7.10. *Let $D(K)$ be a prime, A-adequate diagram with polyhedral decomposition with no non-prime arcs. Then any 2-edge loop in H_A supports at most four EPDs in the spanning set E_c, from at most two color pairs.*

Proof. Any EPD involves a color pair. Recall that Lemma 5.10 on p. 81 implies that for a fixed color pair, at most two EPDs in E_c between those colors run over a given segment. We will show that any brick of H_A can support EPDs in at most two color pairs. Then, the result will follow from Lemma 5.10.

Denote the segments of the 2-edge loop by s_1, s_2. Note that if s_1, s_2 support EPDs of types (1) or (2), then the color pairs of these EPDs are determined by the colors of the tentacles adjacent to s_1 and s_2. For type (3), the colors of adjacent tentacles determine one of the two colors in the pair, and the other is determined by the color of the tentacles meeting the tails of the tentacles of the first color.

Consider the tentacles adjacent to s_1, s_2. There are four such tentacles—one on each side of each segment. There may be two, three, or four distinct colors for these tentacles.

Suppose first there are four distinct colors. Then the 2-edge loop may only support EPDs of type (1) (not (2), not (3)). Since a normal square of type (1) lies on one side of a brick of s_1 and s_2, and since any such brick separates the tentacles into the same pairs inside/outside, there are at most two color pairs in this case.

Now suppose there are three distinctly colored tentacles adjacent to s_1 and s_2. We may have an EPD of type (2) or (3), but not both in distinct color pairs, by Lemma 7.9. If there is an EPD of type (3), then Lemma 7.8 implies there are none of type (1) of distinct color pairs, hence the only EPDs possible are of the same color pair of the EPD of type (3).

If there are three distinctly colored tentacles adjacent to s_1 and s_2, and we have an EPD of type (2), then all color pairs must come from the colors adjacent to the two segments. Label these colors orange, green, and blue, with the colors of the pair of tentacles of the same color labeled orange. Potentially, we might have three color pairs: green–orange, blue–orange, and green–blue. However, note that green and blue tentacles must be on opposite sides of the brick of s_1 and s_2. Since they are distinct colors, only and EPD of type (1) could run through them, but since they are not on the same side of the brick, that is impossible. Thus there are only two color pairs in this case.

Finally, suppose there are only two distinct colors of tentacles adjacent to s_1 and s_2, say green and orange.

If there is an EPD of type (3), then it determines a color pair and there can be no EPD of type (1) with a distinct color pair by Lemma 7.8, nor of type (2) through the same tentacles adjacent to s_1 and s_2 but with a distinct color pair, by Lemma 7.9. There might be an EPD of type (2) and a distinct color pair which uses the other tentacles, but in that case, Lemma 7.9 implies there cannot be another of type (3), hence there are only two possible color pairs.

Similarly, if we have two EPDs of type (3) and distinct color pairs, then they must use distinct tentacles of the brick, and there can be no other types of EPDs with distinct color pairs.

If there is no EPD of type (3), then all EPDs must be of types (1) and (2), for which the colors in the color pair are the colors adjacent to the segments s_1 and s_2. Hence in this case, there is just one color pair.   □

7.3   Estimating the Size of E_c

Now we complete the proof of Theorem 7.2. Lemma 7.10 has the following immediate consequence.

Lemma 7.11. *Let $e_1, \ldots, e_n$ and $f_1, \ldots f_m$ be edges of $\mathbb{G}_A$ (equivalently, segments of the graph H_A), all of which connect the same pair of state circles C and C'. Suppose that the e_i belong to the same twist region, and that the f_j belong to the same twist region. Then the collection of all 2-edge loops of the form $\{e_i, f_j\}$ supports a total of at most four EPDs in the spanning set E_c.*

Proof. Let $\widehat{D}$ be the diagram obtained from D by removing all but one crossing from every A-region of D. Under this operation, $e_1, \ldots, e_n$ become the same edge e of $\mathbb{G}_A(\widehat{D})$, and $f_1, \ldots f_m$ become the same edge f of $\mathbb{G}_A(\widehat{D})$. Furthermore, by Lemma 5.17 on p. 86, there is a one-to-one correspondence between complex disks of $E_c(D)$ and the complex disks of $E_c(\widehat{D})$. In particular, a complex EPD in $E_c(\widehat{D})$ that runs through a tentacle adjacent to the single edge e corresponds to a complex EPD in $E_c(\widehat{D})$ that runs through a tentacle adjacent to one of the e_i, and similarly for the f_j.

Thus, applying Lemma 7.10 to the diagram $\widehat{D}$ gives the desired result for $E_c(D)$.
□

Now Theorem 7.2 will follow immediately from Theorem 5.14 on p. 84 and the following lemma.

Lemma 7.12. *Let E_c be the spanning set of Lemma 5.8 on p. 80, and let m_A be the diagrammatic quantity defined in Definition 7.1 on p. 109. Then, in the absence of non-prime arcs,*

$$0 \leq \|E_c\| \leq 8m_A,$$

with equality if and only if $m_A = 0$.

Proof. Consider a pair of state circles C, C' of $\mathbb{G}_A$, which are connected by at least one edge. There are $e(C, C')$ edges of $\mathbb{G}_A$ connecting these circles, which belong to $m(C, C')$ twist regions. Then the number of bigons between C and C' is $b_A(C, C')$, where

$$b_A(C, C') = e(C, C') - m(C, C'). \tag{7.1}$$

Associated to the pair of circles C and C', we construct a planar surface $S(C, C')$, contained in the projection sphere S^2. Take the disjoint disks in D^2 whose boundaries are C and C', and connect these disks by $m(C, C')$ rectangular bands—with each band containing the segments of the corresponding twist region. Topologically, $S(C, C')$ is a sphere with $m(C, C')$ holes.

Let $D \in E_c$ be an essential product disk that runs through tentacles between C and C'. Then ∂D is a simple closed curve in $S(C, C')$. Now, the conclusion of Lemma 7.11 can be rephrased to say that at most four distinct EPDs of E_c running through tentacles between C and C' can have boundaries that are isotopic in $S(C, C')$. This is because isotopy in $S(C, C')$ is exactly the same equivalence relation as running through tentacles in the same pair of twist regions.

Recall that a sphere with $m(C, C')$ holes contains at most $2m(C, C') - 3$ isotopy classes of disjoint essential simple closed curves. Since the disks in E_c are disjoint, and since at most four of these disks can have boundaries that are isotopic in $S(C, C')$, we conclude that there are at most

$$4(2m(C, C') - 3) < 8(m(C, C') - 1) \tag{7.2}$$

disks in E_c that run through tentacles between C and C'. (Note that if $m(C, C') = 1$, i.e. all segments between C and C' belong to the same twist region, then the left side of (7.2) is negative. But in this case, all EPDs running through tentacles between C and C' must be simple or semi-simple, and cannot belong to E_c. Thus the estimate is meaningful precisely when a 2-edge loop between C and C' contributes to E_c. Meanwhile, the right side of (7.2) is always non-negative when C and C' are connected in $\mathbb{G}_A$.)

Summing over all pairs of state circles C, C' that are connected by at least one edge of $\mathbb{G}_A$, we obtain

$$
\begin{aligned}
\|E_c\| &\leq 8 \sum_{C,C'} (m(C, C') - 1) && \text{by (7.2)} \\
&= 8 \sum_{C,C'} (e(C, C') - b_A(C, C') - 1) && \text{by (7.1)} \\
&= 8 \left(\sum_{C,C'} e(C, C') - \sum_{C,C'} b_A(C, C') - \sum_{C,C'} 1 \right) \\
&= 8 \left(e_A - b_A - e'_A \right) \\
&= 8 m_A && \text{by Def (7.1).}
\end{aligned}
$$

Notice that the inequality is sharp precisely when the estimate of (7.2) applies at least once in a non-trivial way, i.e. when $m_A > 0$. $\qquad\square$

Chapter 8
Montesinos Links

In this chapter, we study state surfaces of Montesinos links, and calculate their guts. Our main result is Theorem 8.6. In that theorem, we show that for every sufficiently complicated Montesinos link K, either K or its mirror image admits an A-adequate diagram D such that the quantity $||E_c||$ of Definition 5.9 vanishes. Then, it will follow that $\chi_-(\text{guts}(M_A)) = \chi_-(\mathbb{G}'_A)$.

8.1 Preliminaries

We begin by reviewing some classically known facts. A reference for this material is, for example, Burde–Zieschang [15, Chap. 12]. A *rational tangle* is a pair (B, L) where B is a 3-ball and L is a pair of arcs in B that are isotopic to ∂B, with the isotopies following disjoint disks in B. Note that a rational tangle is unique up to homeomorphism of pairs. Also, a rational tangle (B, L) contains a unique compression disk that separates the two arcs of L.

A *marked* rational tangle is an embedding of (B, L) into $\mathbb{R}^3$, with B being embedded into the regular neighborhood of a unit square in $\mathbb{R}^2$ (called a *pillowcase*) and the four endpoints of L sent to the four corners of the pillow. For concreteness, we label these four corners NW, NE, SE, and SW. A marked rational tangle specifies a planar projection of K to the unit square, uniquely up to Reidemeister moves in the square.

Marked rational tangles are in 1–1 correspondence with *slopes* in $\mathbb{Q} \cup \{\infty\}$. This can be seen in several ways. First, it is well-known that isotopy classes of essential simple closed curves in a 4-punctured sphere are parametrized by $\mathbb{Q} \cup \{\infty\}$. Thus a rational number determines the slope of a compression disk in the tangle, and this disk determines an embedding of the tangle up to isotopy. A more concrete way to specify the correspondence is to picture the pillowcase boundary of B as constructed from the union of two Euclidean squares. Then, a rational slope q

D. Futer et al., *Guts of Surfaces and the Colored Jones Polynomial*, Lecture Notes in Mathematics 2069, DOI 10.1007/978-3-642-33302-6_8,
© Springer-Verlag Berlin Heidelberg 2013

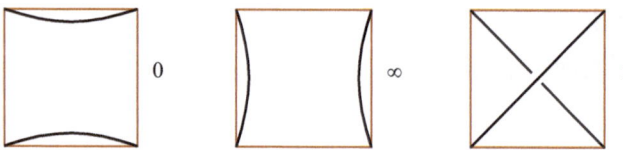

Fig. 8.1 Marked rational tangles of slope 0, ∞, and 1

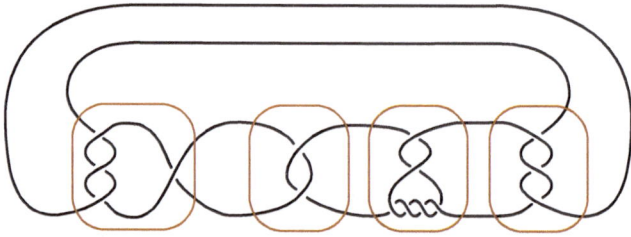

Fig. 8.2 A Montesinos knot constructed from rational tangles of slope $4/3, 1/2, 4/7, -1/3$. This diagram is not reduced, according to Definition 8.3

specifies a Euclidean geodesic that starts from a corner and travels with Euclidean slope q. There are exactly two disjoint arcs with this slope; their union is L.

We adopt the standard convention that rational tangles of slope 0 and ∞ have crossing-free diagrams, and a rational tangle of slope 1 projects to a single positive crossing. See Fig. 8.1.

Given marked rational tangles T_1, T_2 of slope q_1, q_2, one may form a new tangle, called the *sum* of T_1 and T_2, by joining the NE corner of T_1 to the NW corner of T_2 and the SE corner of T_1 to the SW corner of T_2. One may check that if $q_i \in \mathbb{Z}$ for either $i = 1$ or $i = 2$, the result is again a rational tangle of slope $q_1 + q_2$. This is called a *trivial* sum. Otherwise, if $q_i \notin \mathbb{Z}$, the sum tangle will not be rational.

Definition 8.1. For any tangle diagram T with corners labeled NW, NE, SE, and SW, the *numerator closure* of T is defined to be the link diagram obtained by connecting NW to NE and SW to SE by simple arcs with no crossings. The *denominator closure* of T is defined to be the diagram obtained by connecting NW to SW, and NE to SE by simple arcs with no crossings.

Given marked rational tangles $T_1, \ldots, T_r$, a *Montesinos link* is constructed by taking the numerator closure of the sum $T_1 + \ldots + T_r$. We also call this the *cyclic sum* of $T_1, \ldots, T_r$. See Fig. 8.2. In particular, if r is the number of rational tangles used, then the Montesinos link K is determined by an r-tuple of slopes $q_1, \ldots, q_r \in \mathbb{Q} \cup \{\infty\}$. To avoid trivial sums, we always assume $q_i \notin \mathbb{Z}$ for all i.

A cyclic sum of two rational tangles is a two-bridge link. Since two-bridge links are alternating, the guts of the checkerboard surfaces in this case are known by Lackenby's work [58], or equivalently by Corollary 5.19. Thus we assume that $r \geq 3$. In addition, since summing with a tangle of slope ∞ produces a composite or split link, and we are interested in prime links, we also prohibit tangles of slope ∞.

Fig. 8.3 Performing the Euclidean algorithm on p/q produces a continued fraction expansion and an alternating tangle diagram. Here, $3/5 = [0, 1, 1, 2]$

Note, in Fig. 8.2, that a cyclic permutation of the tangles $T_1, \ldots, T_r$ produces the same diagram, up to isotopy in S^2. Furthermore, because every rational tangle admits a rotationally symmetric diagram (Fig. 8.3, left), reversing the order of $T_1, \ldots, T_r$ also does not affect the link. The following theorem of Bonahon and Siebenmann [15, Theorem 12.28] implies that the converse is also true: dihedral permutations of the tangles are essentially the only moves that will produce the same link.

Theorem 8.2 (Theorem 12.28 of [15]). *Let K be a Montesinos link obtained as a cyclic sum of $r \geq 3$ rational tangles whose slopes are $q_1, \ldots, q_r \in \mathbb{Q} \setminus \mathbb{Z}$. Then K is determined up to isomorphism by the rational number $\sum_{i=1}^{r} q_i$ and the vector $((q_1 \mod 1), \ldots, (q_r \mod 1))$, up to dihedral permutation.*

One consequence of Theorem 8.2 is that r, the number of rational tangles used to construct K, is a link invariant. This number is called the *length* of K. In this framework, two-bridge links are Montesinos links of length 2.

For a rational number q, the integer vector $[a_0, a_1, \ldots, a_n]$ is called a *continued fraction expansion* of q if

$$q = a_0 + \cfrac{1}{a_1 + \cfrac{1}{a_2 + \cfrac{1}{\ddots + \cfrac{1}{a_n}}}}.$$

This continued fraction expansion specifies a tangle diagram, as follows. Moving from the outside of the unit square toward the inside, place a_0 positive crossings in a horizontal band, followed by a_1 crossings in a vertical band, etc., until the final a_n crossings in a (vertical or horizontal) band connect all four strands of the braid. The convention is that positive integers correspond to positive crossings and negative integers to negative crossings. The integer a_0 will be 0 if $|q| < 1$, but all subsequent a_i are required to be nonzero. A continued fraction expansion where all nonzero a_j have the same sign as q determines an alternating diagram of the tangle. See Fig. 8.3, where both the alternating diagram and the continued fraction are constructed via a Euclidean algorithm.

Definition 8.3. Let K be a Montesinos link of length $r \geq 3$, obtained as the cyclic sum of tangles $T_1, \ldots, T_r$ of slope $q_1, \ldots, q_r$. A diagram $D(K)$ is called a *reduced Montesinos diagram* if it is a cyclic sum of diagrams of the T_i, and both of the following hold:

(1) Either all q_i have the same sign, or $0 < |q_i| < 1$ for all i.
(2) The diagram of T_i comes from a constant-sign continued fraction expansion of q_i. If the sign is positive, we say T_i is a *positive tangle*. Otherwise, T_i is a *negative tangle*.

Note that $D(K)$ will be an alternating diagram iff all q_i have the same sign.

It is an easy consequence of Theorem 8.2 that every (prime, non-split, non-2-bridge) Montesinos link has a reduced diagram. For example, if $q_i < 0$ while $q_j > 1$, one may add 1 to q_i while subtracting 1 from q_j. By Theorem 8.2, this does not change the link type. Continuing in this fashion will eventually satisfy condition (1) of the definition.

The significance of reduced diagrams lies in the result, due to Lickorish and Thistlethwaite, that for prime, non-split Montesinos links of length $r \geq 3$ the crossing number of K is realized by a reduced diagram [61]. The proof of this result makes extensive use of adequacy. In particular, they make the following observation.

Lemma 8.4 (Lickorish–Thistlethwaite [61]). *Let $D(K)$ be a reduced Montesinos diagram with $r > 0$ positive tangles and $s > 0$ negative tangles. Then $D(K)$ is A-adequate iff $r \geq 2$ and B-adequate iff $s \geq 2$. Since $r + s \geq 3$ in a reduced diagram, D must be either A-adequate or B-adequate.*

Note that if $r = 0$ or $s = 0$, then $D(K)$ is an alternating diagram, which is both A- and B-adequate. Thus it follows from Lemma 8.4 that every Montesinos link is A- or B-adequate. This turns out to be enough to determine the crossing number of K.

In constructing an alternating tangle diagram from a continued fraction, we had a number of choices, as follows. The integer a_1 corresponds to a_1 positive crossings in a vertical band—which can be at the top or bottom of the band. Similarly, the second integer a_2 corresponds to crossings in a horizontal band—which can be at the left or right of the band. For example, in Fig. 8.3, the first crossing was chosen to go on the top of the tangle, and the second crossing was chosen on the left side of the horizontal band. Reversing these choices still produces a reduced diagram. However, for our analysis of I-bundles and guts, we will prefer a particular choice.

Definition 8.5. Let T be a rational tangle of slope q, where $0 < |q| < 1$. If $q > 0$, we say that an alternating diagram $D(T)$ is *admissible* if all the crossings in a vertical band are at the top of the band, and all the crossings in a horizontal band are on the right of the band. If $q < 0$, we say that an alternating diagram $D(T)$ is *admissible* if all the crossings in a vertical band are at the top of the band, and all the crossings in a horizontal band are on the left of the band. For example, the diagram in Fig. 8.3 is admissible.

A reduced Montesinos diagram $D(K)$ is called *admissible* if the sub-diagram $D(T_i)$ is admissible for every tangle of slope $0 < |q_i| < 1$.

For the rest of this chapter, we will assume that $D(K)$ is a reduced, admissible, Montesinos diagram. This assumption does not restrict the class of links under consideration, because every reduced diagram can be made admissible by a sequence of flypes. We also remark that the placement of crossings in vertical and horizontal bands implies that every reduced, admissible diagram is also twist-reduced (see Definition 1.7 on p. 10).

Our goal is to understand the guts of $M_A = S^3 \backslash\backslash S_A$ corresponding to a reduced, admissible diagram.

Theorem 8.6. *Suppose K is a Montesinos link with a reduced admissible diagram $D(K)$ that contains at least three tangles of positive slope. Then D is A-adequate, and*

$$\chi_-(\mathrm{guts}(M_A)) = \chi_-(\mathbb{G}'_A).$$

Note that the A-adequacy of D follows immediately from Lemma 8.4. Similarly, if $D(K)$ contains at least three tangles of negative slope, then it is B-adequate and

$$\chi_-(\mathrm{guts}(M_B)) = \chi_-(\mathbb{G}'_B).$$

Recall that in Theorem 5.14 on p. 84, we have expressed $\chi_-(\mathrm{guts}(M_A))$ in terms of the negative Euler characteristic $\chi_-(\mathbb{G}'_A)$ and the number $||E_c||$ of complex disks required to span the I-bundle of the upper polyhedron. Thus, to prove Theorem 8.6, it will suffice to show that $||E_c|| = 0$. If $D(K)$ is alternating and twist-reduced, all 2-edge loops in $\mathbb{G}_A$ belong to twist regions, hence the result follows by Corollary 5.19. Thus, for the rest of the chapter, we will assume that $D(K)$ is non-alternating; that is, $D(K)$ contains at least three tangles of positive slope and at least one tangle of negative slope.

We will prove the desired statement at the end of the chapter, in Proposition 8.16. In turn, Proposition 8.16 relies on knowing a lot of detailed information about the structure of shaded faces in the upper polyhedron, along with their tentacles. The next section is devoted to compiling this information.

8.2 Polyhedral Decomposition

In this section we will describe the polyhedral decomposition of M_A, in the case of Montesinos diagrams. Then, we prove several tentacle-chasing lemmas about the shape of shaded faces in the upper polyhedron.

Lemma 8.7. *Suppose that a non-alternating diagram $D(K)$ is the cyclic sum of positive slope tangles $P_1, \ldots, P_r$ and negative slope tangles $N_1, \ldots, N_s$. Here, the order of the indices indicates that P_1 is clockwise (i.e. west) of P_2, but does not give*

 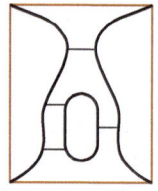 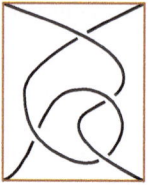 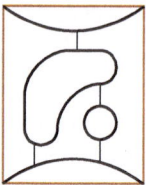

Fig. 8.4 *Left*: A positive tangle and its *A*-resolution. *Right*: A negative tangle and its *A*-resolution

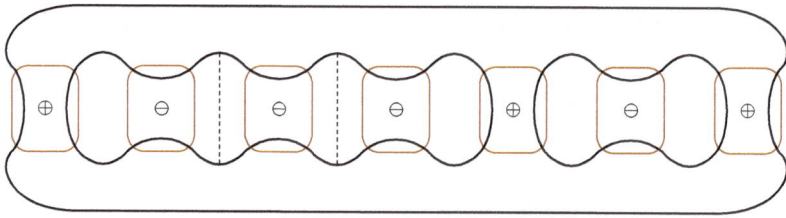

Fig. 8.5 Polyhedral regions of D correspond to individual negative tangles, as well as the sum of all positive tangles. *Dashed lines* are non-prime arcs. A maximal string of consecutive negative tangles, called a *negative block*, is shown on the *left*

any information about the position of P_i relative to any N_j. Then the polyhedral regions of the projection plane are as follows:

(1) *There is one polyhedral region containing all of the positive tangles. The lower polyhedron of this region corresponds to the alternating diagram obtained from a cyclic sum of $P_1, \ldots, P_r$.*

(2) *Every negative tangle N_j corresponds to its own polyhedral region. The lower polyhedron of this region has a diagram coinciding with the alternating diagram of the denominator closure of N_j.*

Recall that the numerator and denominator closures of a tangle, as well as the cyclic sum of several tangles, are defined in Definition 8.1.

Proof. Consider the way in which the state circles of s_A intersect the individual tangles. For each tangle T, its intersection with s_A will contain some number of closed circles, along with exactly two arcs that connect to the four corners of T. When T is a rational tangle, one can easily check that the non-closed arcs of $s_A \cap T$ will run along the north and south sides of T if its slope is positive, and along the east and west sides of T if its slope is negative. See Fig. 8.4.

Now, recall that the polyhedral decomposition described in Chap. 2 proceeds in two steps. In the first step, we cut M_A along white faces; the resulting polyhedra correspond to the (non-trivial) regions in the complement of the state circles s_A. Here, all of the positive tangles are grouped into the same complementary region. On the other hand, every maximal consecutive sequence of negative tangles $N_i, \ldots, N_{i+k}$ defines its own complementary region. See Fig. 8.5. We call such a maximal string of negative tangles a *negative block*.

The second step of the polyhedral decomposition is the cutting along non-prime arcs. The region containing all the positive tangles is prime, and does not need to be cut further. On the other hand, if negative tangles N_i and N_{i+1} are adjacent in D, they will be separated by a non-prime arc α. Thus, at the end of this second step, every negative tangle corresponds to its own lower polyhedron.

The claimed correspondence between these lower polyhedra and alternating link diagrams is a direct consequence of Lemma 2.21 on p. 29.   □

Remark 8.8. Examining Fig. 8.5, along with the shape of individual positive tangles in Fig. 8.4, gives a quick proof of Lemma 8.4.

Lemma 8.9. *Let $D(K)$ be a reduced, admissible, non-alternating, A-adequate Montesinos diagram. Let s be a segment of the graph H_A constructed from D. Consider the two tentacles that run along s. Then*

(1) *At least one of these two tentacles terminates immediately downstream from s.*
(2) *Both tentacles terminate immediately downstream from s, unless s is adjacent to the NE or SW corner of a positive tangle, or the NW or SE corner of a negative tangle.*

Proof. Let C and C' be state circles connected by s. Note that s is contained in some rational tangle T. Suppose, as a warm-up, that that C is an innermost circle entirely contained in T. Then the tentacle that runs downstream toward C will terminate immediately after reaching C, and conclusion (1) is satisfied. We store this observation for later.

To prove the lemma, we consider three cases, conditioned on whether C and/or C' are entirely contained in T.

Case 1: C and C' are entirely contained in T. Then, by the above observation, the tentacles that run downstream to both C and C' terminate immediately. Thus (1) and (2) both hold, proving the lemma in this case.

Case 2: C is entirely contained in T but C' is not. By the above observation, the tentacle that runs downstream toward C will terminate immediately. Also, Fig. 8.4 shows that a state circle C' that is not entirely contained in T must constitute the north or south side of a negative tangle, or else the east or west side of a positive tangle.

Suppose, for example, that T is a positive tangle and that C' forms the east side of it. Then the tentacle running from s toward C' will turn north along C'. Hence, if there is another segment of H_A that meets the east side of T, further north than s, then the tentacle will need to terminate immediately downstream from s. The only way that this tentacle can continue running downstream along C' is if s is adjacent to the NE corner of the tangle.

By the same argument, if T is a positive tangle and C' is the west side of T, then the tentacle that runs downstream toward C' will have to terminate immediately after s, unless s is adjacent to the SW corner of the tangle. Similarly, if T is a negative tangle and C' is its north or south side, then the tentacle that runs downstream toward C' will have to terminate immediately after s, unless s is adjacent to the NW or SE corner of the tangle. This proves the lemma in Case 2.

Case 3: neither C nor C' are contained in T. Then, by the hypotheses of the lemma, these sides of the tangle must be connected by a segment s of H_A. If T is a negative tangle, then the north and south sides of T belong to the same state circle. Thus any segment s spanning T north to south would violate A-adequacy. So T must be a positive tangle, whose west side is C and whose east side is C'.

If s is not adjacent to the NE corner of tangle T, then the argument above implies that the tentacle running downstream toward C' will terminate immediately downstream from s. Similarly, the tentacle running downstream toward C will terminate, unless s is adjacent to the SW corner of T. The only situation in which neither of these tentacles terminate immediately after s is the one where s is simultaneously adjacent to the NE and SW corners of the tangle. But then the tangle T contains a single crossing, and has slope $+1$, violating Definition 8.3 of a reduced diagram. This proves the lemma.   □

Lemma 8.10. *Let $D(K)$ be as in Lemma 8.9 and let γ be a simple arc in a shaded face that starts in an innermost disk of H_A. (See Definition 3.2 on p. 37.) Then the course of γ must satisfy one of the following:*

(1) *γ terminates after running downstream along 2 or fewer segments of H_A.*
(2) *γ terminates after running downstream along 3 segments of H_A, where the middle segment spans a positive tangle east to west.*
(3) *γ runs downstream along one tentacle, through a non-prime switch, and then upstream along one tentacle into another innermost disk. In this case, both innermost disks belong to consecutive negative tangles.*

See Fig. 3.2 on p. 36 for a review of tentacles and non-prime switches. One way to summarize the conclusion of the lemma is that, in the special case of admissible Montesinos diagrams, right-down staircases in H_A have at most 3 stairs.

Proof. By hypothesis, γ starts in an innermost disk of H_A. Let s_1 be the segment of H_A along which γ runs out of this innermost disk. If s_1 is not adjacent to the NE or SW corner of a positive tangle, or the NW or SE corner of a negative tangle, then Lemma 8.9 implies γ must terminate immediately, and conclusion (1) holds.

Suppose s_1 belongs to a positive tangle T_1, and is adjacent to its NE corner. By Lemma 8.7, the polyhedral region containing T_1 does not have any non-prime arcs, hence γ cannot enter a non-prime switch. Thus the only way γ can continue running downstream is by entering a negative tangle T_2 along a segment s_2 on the north side of T_2. As we have already seen, a negative tangle cannot be spanned north to south by a single segment (otherwise, it would have integer slope and violate Definition 8.5). Thus s_2 connects to an innermost circle, and γ must terminate after two segments.

Similarly, if s_1 is adjacent to the SW corner of a positive tangle T_1, then γ can only continue running downstream by entering a negative tangle T_2 along a segment s_2 in on the south side of T_2. After this, γ must terminate after two segments, and conclusion (1) holds.

Suppose s_1 belongs to a negative tangle T_1, and is adjacent to its SE corner. After running along s_1, γ can enter a non-prime switch, or continue running downstream through tentacles. Suppose, first, that γ enters a non-prime switch. All the non-prime arcs of $D(K)$ separate consecutive negative tangles, as in Fig. 8.5. After entering a non-prime switch from the SE corner of T_1, γ can run downstream to the NE corner of T_1 and terminate, run downstream to the SW corner of the next negative tangle T_2 and terminate, or run upstream to the NW corner of the negative tangle T_2. After this, γ is forced to run upstream into an innermost disk, as in conclusion (3).

Next, suppose that after running along s_1, γ continues downstream through tentacles. Thus γ must enter a positive tangle T_2 through a segment s_2 on the west side of T_2. If s_2 leads to an innermost disk, then γ must terminate after two steps. Alternately, if s_2 spans the positive tangle T_2 east to west but is not adjacent to the NE corner of T_2, then γ must also terminate after two steps. Finally, if s_2 spans the positive tangle T_2 east to west and is adjacent to the NE corner of T_2, then we can repeat the analysis of positive tangles at the beginning of the proof (with s_2 playing the role of s_1). In this case, γ must terminate after at most three steps. If there are indeed 3 steps, then the middle segment s_2 spans a positive tangle from west to east, and conclusion (2) holds.

Finally, suppose s_1 belongs to a negative tangle T_1, and is adjacent to its NW corner. Then we argue as above, with all the compass directions reversed, and reach the same conclusions.   □

Lemma 8.11 (Head locator). *Let $D(K)$ be a reduced, admissible, A-adequate Montesinos diagram that is non-alternating. If a shaded face in the upper polyhedron of the polyhedral decomposition of M_A meets an innermost disk (has a head) in a positive tangle, then it meets no other innermost disks elsewhere. If a shaded face meets an innermost disk in a negative tangle, then it may meet another innermost disk in the same negative block, but it will not meet any innermost disk in any other negative block.*

See Fig. 8.5 for the notion of a negative block. Recall as well that a *head* of a shaded face is an innermost disk, as in Fig. 3.1 on p. 36.

Proof. A shaded face meets more than one innermost disk only when it runs through a non-prime switch. Since all non-prime arcs in our diagrams lie inside of negative blocks, only innermost disks inside the same negative block can belong to the same shaded face.   □

Lemma 8.12. *Let $D(K)$ be a reduced, admissible, non-alternating Montesinos diagram with at least three positive tangles. Then tentacles of the same shaded face in the upper polyhedron of the polyhedral decomposition of M_A cannot run across the north and the south of the outside of a single negative block.*

Proof. Note that by Lemma 8.4 $D(K)$ is A-adequate. The tentacle across the north of the outside of a negative block runs from a segment in the positive tangle directly to the west of that negative block. See Fig. 8.5. Hence, by Lemma 8.11 (Head locator), its head is either inside that positive tangle, or, if it came from a segment running east to west in that positive tangle, its head will be inside the negative block directly to the west.

Similarly, the tentacle across the south of the outside of a negative block runs from a segment in the negative tangle directly to the east of that negative block, hence has its head inside the positive tangle directly to the east, or the negative block directly to the east.

Note that the positive tangles directly to the east and west cannot agree, by the assumption that our diagram has at least three positive tangles. Similarly, the negative blocks to the east or west cannot agree. Hence the conclusion follows. □

Lemma 8.13. *Suppose that $D(K)$ is a reduced, admissible Montesinos diagram with at least three positive tangles that is not alternating. Any tentacle in the polyhedral decomposition of M_A running over the outside of a negative block cannot have a head inside that negative block.*

Proof. The head of a tentacle running over the outside of a negative block lies in the positive tangle or the negative block directly to the west, in case the tentacle runs across the north, or in the positive tangle or negative block directly to the east, in case the tentacle runs across the south. Then the result follows from Lemma 8.11 (Head locator).   □

8.3 Two-Edge Loops and Essential Product Disks

Next, we study EPDs in the upper polyhedron of the polyhedral decomposition of M_A. Recall that, by Corollary 6.6, every EPD in the upper polyhedron P must run through tentacles adjacent to a 2-edge loop in $\mathbb{G}_A$. In Lemma 8.14 below, we show that these 2-edge loops can be classified into three different types. Most of our attention will be devoted to particular type of 2-edge loop, depicted in Fig. 8.6.

Lemma 8.14. *Let $D(K)$ be a reduced, admissible, A-adequate, non-alternating Montesinos diagram. Let C, C' be a pair of state circles of s_A. These circles are connected by multiple segments of H_A (corresponding to a two-edge loop of $\mathbb{G}_A$) if and only if one of the following happens:*

(1) *C and C' co-bound one or more bigons in the short resolution of a twist region, which is entirely contained in a tangle. See Fig. 5.4.*
(2) *C is contained inside a negative tangle N_i of slope $-1 < q \leq -1/2$, and is connected by segments of H_A to the state circle C' that runs along the north and south of N_i. See Fig. 8.6.*
(3) *There are exactly two positive tangles P_1 and P_2, and C, C' are the state circles that run along the east and west sides of these tangles.*

Proof. If the circles C, C' satisfy one of the conditions of the lemma, it is easy to see that they will be connected by two or more segments of H_A. To prove the converse, suppose that C and C' are connected by two segments of H_A. Each of these segments corresponds to a crossing of the diagram D, and belongs to a

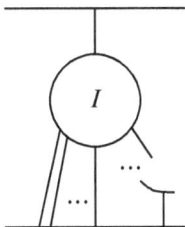

Fig. 8.6 The form of a negative tangle with a 2-edge loop of type (2) of Lemma 8.14. There is exactly one segment at the north, and one or more parallel segments on the south–west. The south–east of the tangle may or may not have additional segments and state circles. The segment on the north of innermost disk I, plus a segment on the south I, forms a *bridge* from the north to the south of the tangle

particular rational tangle. We consider several cases, conditioned on how C and C' intersect this common tangle.

First, suppose that C and C' are both closed loops inside a tangle T. Since the tangle diagram is alternating, each of these state circles is innermost. Thus there is a loop in the projection plane that runs through the regions bounded by C and C', and intersects the projection of K exactly at the two crossings where C meets C'. Then these crossings are twist-equivalent. Since a reduced, admissible Montesinos diagram must be twist-reduced, the two crossings are connected by one or more bigons. Thus conclusion (1) holds.

Next, suppose that C is entirely contained in a tangle T, and that $C' \cap T$ consists of one or two arcs. If the two segments of H_A connect C to the same arc on the side of a tangle, then again the corresponding two crossings are twist-equivalent, and must be connected by one or more bigons. Thus conclusion (1) holds. If the two segments of H_A connect C to opposite sides of the tangle, then C' must contain both of those arcs (east and west in the case of a positive tangle, north and south in the case of a negative tangle). We investigate this possibility further.

If $T = P_i$ is a positive tangle, then Fig. 8.5 shows that the state circle on the east side of P_i also runs along the west side P_{i+1}, but is disjoint from every other positive tangle. Thus the east and west sides of P_i belong to the same state circle only if there is exactly one positive tangle—but this violates A-adequacy, by Lemma 8.4.

If T is a negative tangle, then Fig. 8.5 shows that the north and south sides of T will indeed belong to the same state circle C'. Now, let $q < 0$ be the slope of T, and let $[a_0, a_1, \ldots, a_n]$ be the continued fraction expansion of q. Since the diagram D is reduced, we have $q \in (-1, 0)$, hence $a_0 = 0$. Moving from the boundary of the tangle inward, the first crossings will be $|a_1|$ negative crossings in a vertical band. If $|a_1| \geq 3$, or if $|a_1| = 2$ and $|a_2| > 0$, then the north and south sides of T will not connect to the same state circle in T. Note that these conditions on a_1 and a_2 are describing exactly the rational numbers $|q| < 1/2$. On the other hand, if $-1 < q \leq -1/2$, there is exactly one circle C that connects to the north and south sides of the tangle. This state circle is the boundary of an innermost disk I depicted in Fig. 8.6. Thus conclusion (2) holds.

Finally, suppose that neither C nor C' is entirely contained in a single tangle. Then each segment of H_A that connects C to C' must span all the way from the north to the south sides of a tangle, or from the east to the west. Since the diagram $D(K)$ is reduced and non-alternating, every tangle has slope $|q_i| < 1$, hence no single edge of H_A can span a tangle from north to south. The remaining possibility is that each segment of H_A that connects C to C' spans a positive tangle east to west. If these two segments lie in the same tangle T, the corresponding crossings are twist-equivalent, hence conclusion (1) holds. If these two segments lie in two different positive tangles P_1 and P_2, and the east and west sides of these tangles belong to the same state circles C, C', then P_1 and P_2 must be the only positive tangles in the diagram. Thus conclusion (3) holds.   □

Lemma 8.14, combined with Corollary 6.6, will allow us to find and classify EPDs in the polyhedral decomposition for Montesinos links. Looking over the conclusions of Lemma 8.14, we find that two-edge loops of type (1) are very standard, and easy to deal with using Lemma 5.17. Loops of type (3) will be ruled out once we assume that $D(K)$ has at least three positive tangles. Thus most of our effort is devoted to studying EPDs that run over a two-edge loop of type (2).

It is worth taking a closer look at negative tangles that support a two-edge loop of (2). The A-resolution of such a negative tangle is illustrated in Fig. 8.6. Note in particular that there is exactly one segment connecting the innermost disk I to the outside of the negative block at the north of the tangle. On the south, one or more parallel segments connects the innermost disk to the outside of the negative block on the south, and these segments are all at the south–west of the diagram. The portion of the diagram on the south–east can have additional state circles and edges, or not. The 2-edge loop runs over the single segment in the north and one of the parallel strands in the south–west.

Lemma 8.15. *Let N_i be a negative tangle in a reduced, A-adequate, Montesinos diagram $D(K)$, with at least three positive tangles. Let E be an essential product disk in the upper polyhedron, which runs over a 2-edge loop of that spans N_i north to south, as in Fig. 8.6. Then ∂E must run through the innermost disk I shown in Fig. 8.6. Furthermore, ∂E must run adjacent to the 2-edge loop through at least one tentacle whose head is the innermost disk I.*

Proof. By Lemma 8.9, all tentacles that run downstream toward I must terminate upon reaching I. Thus, each time ∂E runs along a segment that connects I to the north or south sides of the tangle, it must either pass through the tentacle that runs downstream out of I (hence, through the innermost disk I), or through a tentacle that runs downstream toward I (hence, into I). In either case, the disk E must run through I.

Next, suppose for a contradiction that on both the north and south sides of the tangle, ∂E runs in tentacles that terminate at I. Since ∂E intersects only two shaded faces, one of which has a head at I, the two tentacles running toward I from the north and south must belong to the same shaded face. But then the heads of these two tentacles are both attached to the outside of the negative block, one on the north and one on the south, and so the negative block must have tentacles of this same color both on the north and on the south. This contradicts Lemma 8.12. Therefore, ∂E must run downstream out of I, either on the north or on the south (or both).   □

8.4   Excluding Complex Disks

We are now ready to prove the following proposition. As remarked earlier, the statement that $||E_c|| = 0$ and the upper polyhedron contains no complex disks, combined with Theorem 5.14, immediately implies the non-alternating case of Theorem 8.6.

Proposition 8.16. *Suppose that $D(K)$ is a reduced, admissible, non-alternating Montesinos link diagram with at least three positive tangles. Then every EPD in the upper polyhedron is either parallel to a white bigon face (simple), or parabolically compresses to bigon faces (semi-simple). In particular, in the terminology of Lemma 5.8, $E_c = \emptyset$.*

Proof. Let E be an essential product disk in the upper polyhedron. By Corollary 6.6, ∂E must run over tentacles adjacent to a 2-edge loop in $\mathbb{G}_A$. With the hypothesis that $D(K)$ has at least three positive tangles, Lemma 8.14 implies that every 2-edge loop in $\mathbb{G}_A$ is of type (1) or (2).

Type (1) loops correspond to crossings in a single twist region, in which the all-A resolution is the short resolution (see Fig. 5.4 on p. 86). Note that by Lemma 5.17, removing all the bigons in the A-twist regions does not affect the number $||E_c||$ of complex disks in the spanning set of the upper polyhedron. But if all bigons in A-regions are removed, the only remaining two-edge loops will be of type (2), spanning a negative tangle north to south. Thus, if we can show that every EPD running over a type (2) loop is simple or semi-simple, it will follow that the same conclusion holds for type (1) loops as well.

For the remainder of the proof, we assume that E is an essential product disk, such that ∂E runs over tentacles adjacent to a 2-edge loop that spans a negative tangle north to south, as in Fig. 8.6. Then, by Lemma 8.15, ∂E must run through an innermost disk I, as in Fig. 8.6. We will show that E is either parallel to a bigon face, or parabolically compressible to a collection of bigon faces.

Following the setup of Chap. 6, color the shaded faces met by E orange and green,[1] so that the shaded face containing the innermost disk I is green. As in Lemma 6.1 (EPD to oriented square), we may pull ∂E off the shaded faces, forming a normal square. Note that under the orientation convention of Lemma 6.1 (EPD to oriented square), any arc of the normal square on a white face cuts off a vertex at the head of a green tentacle and at the tail of an orange tentacle.

---

[1] Note: For versions of the monograph in grayscale, orange faces will appear in the figures as light gray, green as darker gray.

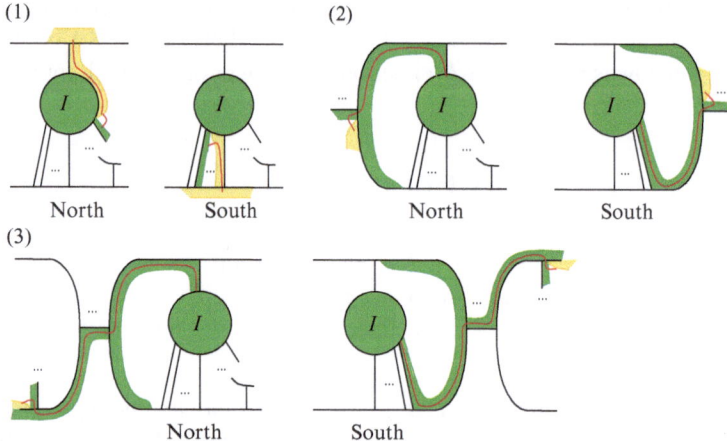

Fig. 8.7 Ideal vertices of an EPD, of types (1), (2), and (3). In each type, the boundary of the EPD, shown in *red*, and can run north or south from the innermost disk. Both cases are shown

We consider the possible locations of the ideal vertices of E. Equivalently, we consider the possible white faces into which ∂E has been pulled. Using Lemma 8.10, we may enumerate the possible locations for an ideal vertex of E. These are shown in Figs. 8.7–8.9.

(1) An ideal vertex of ∂E may appear on the innermost disk of Fig. 8.6 itself. This means the vertex is at the head of a tentacle running north or south out of the innermost disk I.

(2) An ideal vertex of ∂E may appear at the head of a tentacle running out of the negative block containing I, if ∂E runs downstream from I to the next adjacent positive tangle.

(3) If ∂E runs downstream from I to the next adjacent positive tangle, then across a segment spanning the positive tangle east to west, and then downstream across the outer state circle of the next negative block, the vertex may appear on the next adjacent negative block.

(4) If ∂E runs downstream from I, across a non-prime arc, and then upstream, then it will run through an innermost disk in the next adjacent negative tangle. The vertex may appear on this innermost disk. Note the corresponding orange tentacle will either come from an innermost disk inside the negative tangle, or from a tentacle across the north or south of the negative block. All three of these possibilities are shown in Fig. 8.8.

(5) If ∂E runs across a non-prime arc, then upstream into an innermost disk J, and the vertex does not appear on this innermost disk, then ∂E must run downstream again from J. In fact, to obtain the correct orientation near an ideal vertex of E, its boundary must run downstream for at least two stairs in a staircase starting at J. Thus, by Lemma 8.10, ∂E runs along a second 2-edge loop, with an innermost disk at J. (See Fig. 8.9.) Exiting this 2-edge loop, one of the vertices

Fig. 8.8 Ideal vertices of an EPD, of type (4). Note the boundary of the EPD, shown in *red*, can run north or south. The *orange* tentacle it meets can either come from an innermost disk, from the south, or from the north of the negative block. All possibilities are shown

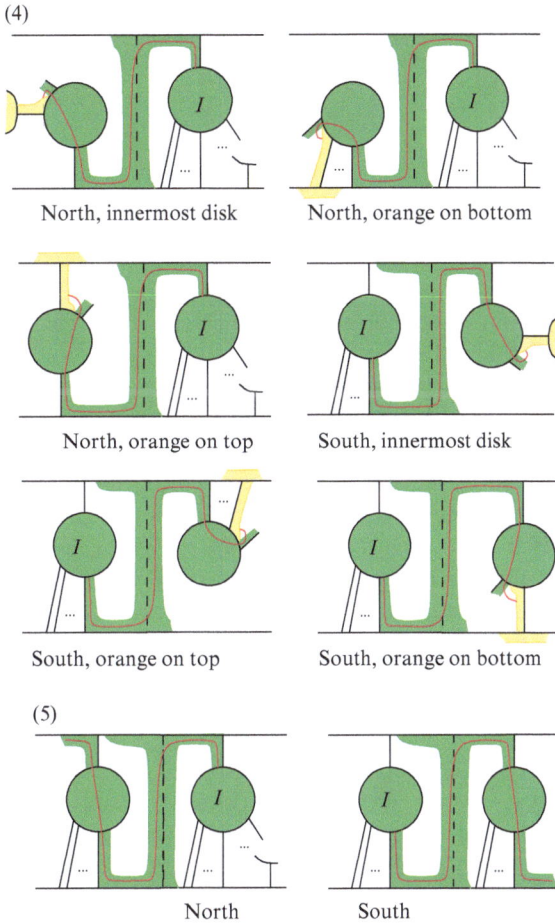

(4)

North, innermost disk

North, orange on bottom

North, orange on top

South, innermost disk

South, orange on top

South, orange on bottom

Fig. 8.9 Ideal vertices of an EPD, of type (5). The boundary of the EPD, shown in *red*, can run north or south. Both cases are shown

(5)

North

South

of types (2), (3), (4), or (5) must occur. Note that between the pairs of 2-edge loops is a collection of bigons. We will handle this last type by induction on the number of negative tangles in a negative block.

Now, each EPD has two vertices. From the green innermost disk I, ∂E runs through the green shaded face in two directions (north and south) toward these two vertices. Each must be one of the above enumerated types. We consider all the combinations of these types of vertices.

Type (1) and type (1): This combination cannot happen. For if both vertices lie on the given innermost disk, then ∂E does not run through any green tentacles exiting the innermost disk, which contradicts Lemma 8.15.

Type (1) and type (2): In this case, type (2) implies the negative tangle containing the innermost disk I is to the far west or far east of the negative block, and the green tentacle leaves this negative tangle and wraps around to the side of the

Fig. 8.10 Combining
vertices of type (1) and (2).
Left: type (1) north and (2)
south; *right*: type (1) south
and (2) north

(1) and (2)

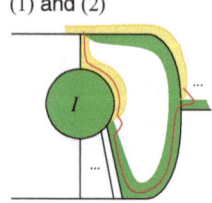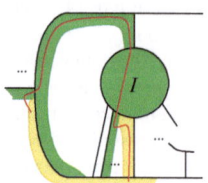

adjacent positive tangle. Note that although ∂E meets a vertex here, the tentacle itself must continue until it terminates at the same negative tangle, forming a white bigon face. See Fig. 8.10.

Next, note that since one of the vertices is of type (1), a tentacle across the outside of the given negative block must be orange (either top or bottom, depending on whether the EPD runs through an orange tentacle of the 2-edge loop on the north or south). Since the other vertex, which lies on the outside of the negative block, must also meet an orange tentacle, Lemma 8.11 (Head Locator) implies the only possibility is that the other vertex meets the same orange tentacle, and ∂E runs through this orange tentacle connecting the vertices. In this case, ∂E bounds only bigon(s), and E is parabolically compressible to bigons.

Type (1) and type (3): This case cannot occur by the Head Locator Lemma 8.11; for, the head of the orange in type (1) would lie in the positive tangle or negative block adjacent to one side, and the head of the orange in type (3) would lie in the negative block to the opposite side.

Type (2) and type (2): In this case, tentacles from the north and south of the negative tangle run along the sides of the positive tangles to the east and west of the negative block. Lemma 8.11 (Head Locator) puts serious restrictions on the diagram from here. (See Fig. 8.11.) In particular, by Lemma 8.11, any head(s) of the orange face must lie in the same positive tangle, or in the same negative block. Note that when the vertex on the west of the negative block is of this type, then the orange tentacle it meets must either

(a) Run across the south of the given negative block, if the vertex is at the very south-east of the positive tangle, and have its head in the positive tangle or negative block to the east, or
(b) Come from an innermost disk in that positive tangle to the west of our negative block, or
(c) Come from an innermost disk inside the negative block to the west of our original negative block.

Three similar options hold for the east. Lemma 8.11 (Head locator) implies that only one of two possibilities may occur: on the west, the vertex lies at the south–east tip of the positive tangle, and the orange head(s) are in the positive tangle or negative block to the east, or on the east the vertex lies at the north–west tip of the positive tangle, and all orange head(s) are in the positive tangle or negative block to the west.

(2) and (2)

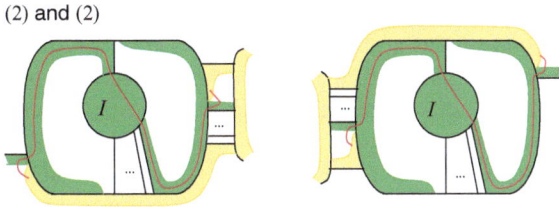

Fig. 8.11 Possibilities for type (2) and (2)

(2) and (3)

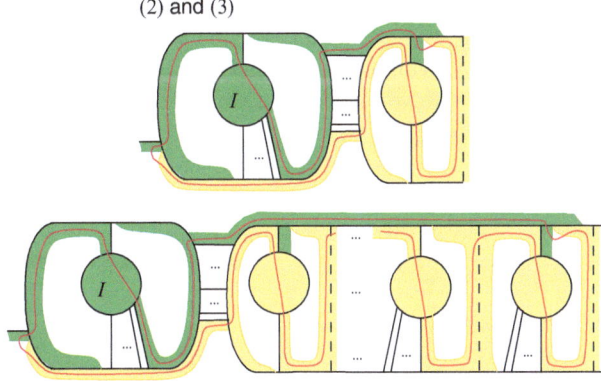

Fig. 8.12 Possibilities for types (2) north and (3) south. Note there are similar possibilities for (2) south and (3) north

We argue that E parabolically compresses to bigons. Both cases are similar; we go through the case that the head of the orange lies to the east (Fig. 8.11, left).

First, note that since each innermost disk of a positive tangle has a distinct color, the two segments attached to orange tentacles that meet vertices of ∂E must have a head in the same orange innermost disk in the positive tangle. Thus the segment at the south–west of the positive tangle to the east, which must have orange on one side, must be attached to the same state circles as the segment near the vertex on the east of the negative block. Then these segments are twist-equivalent, hence bound a chain of bigons Hence, ∂E must run from the vertex on the west of the negative block, through the tentacle across the south, up the segment at the south–west of the positive tangle to the east, then encircle bigons, and connect to the vertex on the east of the negative block. The diagram must be as shown in Fig. 8.11, and the EPD is parabolically compressible to bigons.

Type (2) and type (3): The appearance of a type (3) vertex forces the head(s) of the orange shaded face to lie in the negative block adjacent to one side, which, just as above in the case (2) and (2), puts restrictions on the portion of the diagram with the vertex of type (2). The argument is similar if the orange innermost lies to the east or to the west. For ease of explanation, we go through the case that the orange lies to the east. The result is illustrated in Fig. 8.12.

In particular, in this case the orange tentacle of vertex (2) must run across the south of the outside of the negative block, and across a segment spanning the next (eastward) positive tangle from east to west. Further, there must be an orange tentacle just inside the adjacent negative block to the east, adjacent to this positive tangle. This tentacle comes from an innermost disk J in the first negative tangle in this negative block. The other vertex, of type (3) meets a tail of another orange tentacle inside this negative block, at the top. Thus, the innermost disk at the head of this other tentacle either agrees with J, or is connected to J by a sequence of non-prime switches. If J is the head of both tentacles, then they must bound bigons between them, and E parabolically compresses to bigons, as in Fig. 8.12. If J is not the head of both tentacles, then there is a sequence of 2-edge loops of the type shown in Fig. 8.9, this time with an orange innermost disk, and E running across each loop in a string separated by non-prime arcs. Again this bounds bigons, and parabolically compresses to bigons.

Type (3) and type (3): This case cannot occur. Two vertices of this type would require orange heads in both negative blocks to the east and west of the given negative block, contradicting Lemma 8.11 (Head locator).

Next, consider vertices of type (4): the vertex lies on an innermost disk in the adjacent negative tangle. Again we analyze the possible locations for orange heads from the other vertex, using Lemma 8.11 (Head Locator).

Type (1) and type (4): In this case, an orange tentacle meeting a vertex of type (1) must run over the outside of the negative block, so Lemmas 8.12 and 8.13 imply that the innermost disk in the adjacent negative tangle must meet an orange tentacle connected to the same side of the outside of the given negative block. This implies ∂E will bound a sequence of bigons, and hence E parabolically compresses to bigons. See Fig. 8.13, left.

Type (2) and type (4): Here, the vertex of type (2) meets an orange tentacle on the outside of the negative block. Thus Lemmas 8.12 and 8.13 imply that the orange tentacle meeting the vertex of type (4) must have its head attached to the outside of the negative block. Then, ∂E must run through this outside tentacle toward the vertex of type (2). If the tentacle terminates at the vertex of type (2), then ∂E encloses only bigons. Otherwise, the tentacle has its head attached to another state circle. In this case, the argument is the same as the one in case (2) and (2) above. The orange tentacle meeting the other vertex is also attached to this state circle, and we have a parabolic compression to bigons.

Type (3) and type (4): In this case, the head of an orange tentacle meeting the vertex of type (3) must be in the next adjacent negative block. Thus, by the Head Locator Lemma 8.11, the orange tentacle meeting the vertex of type (4) must have its head inside the same adjacent negative block. Then, an argument similar to that in case (2) and (3) implies that ∂E encircles only bigons. Compare Fig. 8.12 to Fig. 8.13.

Type (4) and type (4): Here, Lemmas 8.12, and 8.13 imply that the orange tentacles meeting the two vertices cannot come from the north on one side and the south on the other, or the north or south on one side and an innermost disk on the other. Neither can both orange tentacles come from innermost disks in distinct

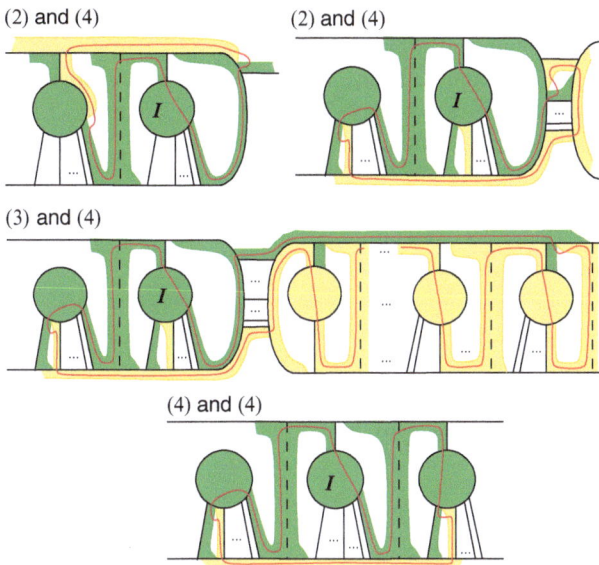

Fig. 8.13 All possibilities for type (4) north

negative tangles, for those tangles will be separated by green non-prime switches, not orange connecting switches. The only remaining possibility is that the orange tentacles come from outside the negative block on the same side. In this case, ∂E bounds bigons, as desired. See Fig. 8.13.

Type (5) cases: Recall that in type (5), an arc of ∂E runs over a non-prime arc and upstream, but the next vertex does not lie on an innermost disk in this negative tangle. Then ∂E must continue downstream out of this innermost disk. Thus we have another 2-edge loop as in Fig. 8.6, and the options from Lemma 8.10 imply that from here, ∂E cannot meet a vertex immediately, so its path toward a vertex is one of type (2), (3), (4), or (5). By induction on the number of negative tangles in a negative block, there will be some finite number of consecutive 2-edge loops corresponding to instances of case (5), but eventually ∂E will run to a vertex of types (2), (3), or (4). Note that between these 2-edge loops from type (5), we have only bigon faces. Combining the above arguments with these additional bigon faces, we find that in all cases ∂E encloses only bigons.

This phenomenon is illustrated in the bottom panel in Fig. 8.13. Thinking of the middle green innermost disk I as the innermost disk of the representative 2-edge loop, we argued that this figure arose by combining vertices of type (4) and (4). However, if we think of the right-most innermost green disk as the innermost disk of our representative 2-edge loop, then this figure illustrates a vertex of type (1) (bottom right), from which ∂E runs over a second 2-edge loop to the west, which is type (5), followed by a vertex of type (4). More generally, we could have n negative tangles as in the middle of the bottom panel of Fig. 8.13, strung end to end. Between such tangles, ∂E bounds only bigons.

This completes the enumeration of cases. For every combination of ideal vertices, E parabolically compresses to bigons. Thus there are no complex EPDs in the upper polyhedron. This completes the proof of Proposition 8.16, hence also the proof of Theorem 8.6.   □

Chapter 9
Applications

In this chapter, we will use the calculations of $\mathrm{guts}(S^3 \backslash\backslash S_A)$ obtained in earlier chapters to relate the geometry of A-adequate links to diagrammatic quantities and to Jones polynomials. In Sect. 9.1, we combine Theorem 5.14 with results of Agol, Storm, and Thurston [6] to obtain bounds on the volumes of hyperbolic A-adequate links. A sample result is Theorem 9.7, which gives tight diagrammatic estimates on the volumes of positive braids with at least 3 crossings per twist region. The gap between the upper and lower bounds on volume is a factor of about 4.15.

In Sect. 9.2, we apply these ideas to Montesinos links, and obtain diagrammatic estimates for the volume of those links. Again, the bounds are fairly tight, with a factor of 8 between the upper and lower bounds.

In Sect. 9.3, we relate the quantity $\chi_-(\mathrm{guts}(S^3 \backslash\backslash S_A))$ to coefficients of the Jones and colored Jones polynomials of the link $K = \partial S_A$. One sample application here is Corollary 9.16: for A-adequate links, the next-to-last coefficient β_K' detects whether a state surface is a fiber in $S^3 \setminus K$. Finally, in Sect. 9.4, we synthesize these ideas to obtain relations between the Jones polynomial and volume. As a result, the volumes of both positive braids and Montesinos links can be bounded above and below in terms of these coefficients.

9.1 Volume Bounds for Hyperbolic Links

Using Perelman's estimates for volume change under Ricci flow with surgery, Agol, Storm, and Thurston [6] have obtained a relationship between the guts of an essential surface $S \subset M$ and the hyperbolic volume of the ambient 3-manifold M. The following result is an immediate consequence of [6, Theorem 9.1], combined with work of Miyamoto [68, Proposition 1.1 and Lemma 4.1].

D. Futer et al., *Guts of Surfaces and the Colored Jones Polynomial*, Lecture Notes in Mathematics 2069, DOI 10.1007/978-3-642-33302-6_9,
© Springer-Verlag Berlin Heidelberg 2013

Theorem 9.1. *Let M be finite-volume hyperbolic 3-manifold, and let $S \subset M$ be a properly embedded essential surface. Then*

$$\mathrm{vol}(M) \geq v_8 \, \chi_-(\mathrm{guts}(M \backslash\backslash S)),$$

where $v_8 = 3.6638\ldots$ is the volume of a regular ideal octahedron.

Remark 9.2. By [6] and work of Calegari, Freedman, and Walker [16], the inequality of Theorem 9.1 is an equality precisely when S is *totally geodesic* and $M \backslash\backslash S$ is a union of regular ideal octahedra. We will not need this stronger statement.

In general, it is hard to effectively compute the quantity $\chi_-(\mathrm{guts}(M \backslash\backslash S))$ for infinitely many pairs (M, S). To date, there have only been a handful of results computing the guts of essential surfaces in an infinite family of manifolds: see e.g. [3, 57, 58]. In particular, Lackenby's computation of the guts of checkerboard surfaces of alternating links [58, Theorem 5] enabled him to estimate the volumes of these link complements directly from a diagram. See [58, Theorem 1] and [6, Theorem 2.2].

In the A-adequate setting, we have the following volume estimate.

Theorem 9.3. *Let $D = D(K)$ be a prime A-adequate diagram of a hyperbolic link K. Then*
$$\mathrm{vol}(S^3 \setminus K) \geq v_8 \, (\chi_-(\mathbb{G}'_A) - \|E_c\|),$$
where $\chi_-(\mathbb{G}'_A)$ and $\|E_c\|$ are as in the statement of Theorem 5.14 and $v_8 = 3.6638\ldots$ is the volume of a regular ideal octahedron.

Proof. We will apply Theorem 9.1 to the essential surface S_A and the 3-manifold $S^3 \setminus K$. Since $S^3 \backslash\backslash S_A$ is homeomorphic to $(S^3 \setminus K) \backslash\backslash S_A$, we have

$$\mathrm{vol}(S^3 \setminus K) \geq v_8 \, \chi_-(\mathrm{guts}(S^3 \backslash\backslash S_A)) = v_8 \, (\chi_-(\mathbb{G}'_A) - \|E_c\|),$$

where the equality comes from Theorem 5.14. The result now follows.   □

Theorem 9.3 becomes particularly effective in the case where $\|E_c\| = 0$. For example, this will happen when every 2-edge loop in the state graph $\mathbb{G}_A$ comes from a single twist region of the diagram D.

Corollary 9.4. *Let $D(K)$ be a prime, A-adequate diagram of a hyperbolic link K, such that for each 2-edge loop in $\mathbb{G}_A$, the edges belong to the same twist region of $D(K)$. Then*
$$\mathrm{vol}(S^3 \setminus K) \geq v_8 \, (\chi_-(\mathbb{G}'_A)).$$

Proof. This follows immediately from Theorem 9.3 and Corollary 5.19.   □

Remark 9.5. If $D = D(K)$ is a prime reduced alternating link diagram, then the hypotheses of Corollary 9.4 are satisfied by both the state graphs $\mathbb{G}_A$ and $\mathbb{G}_B$. Thus Corollary 9.4 gives lower bounds on volume in terms of both $\chi_-(\mathbb{G}'_A)$ and $\chi_-(\mathbb{G}'_B)$. By averaging these two lower bounds, one recovers Lackenby's lower bound on

Fig. 9.1 The generators σ_1 and σ_2 of the 3-string braid group

the volume of hyperbolic alternating links, in terms of the twist number $t(D)$ [6, Theorem 2.2].

Corollary 9.4 also applies to certain closed braids.

Definition 9.6. Let B_n denote the braid group on n strings. The elementary braid generators are denoted $\sigma_1, \ldots, \sigma_{n-1}$ (see Fig. 9.1 for the case $n = 3$). A braid $b = \sigma_{i_1}^{r_1} \sigma_{i_2}^{r_2} \cdots \sigma_{i_k}^{r_k}$ is called *positive* if all the exponents r_j are positive, and *negative* if all the exponents r_j are negative.

Suppose that D_b is the closure of a positive braid $b \in B_n$. Then it follows immediately that the diagram D_b is B-adequate. In fact, the reduced graph $\mathbb{G}'_B$ is a line segment with n vertices. Thus, by Theorem 5.11, the state surface S_B is a fiber for $S^3 \setminus K$. (This recovers a classical result of Stallings [89] and Gabai [40].) In particular, $S^3 \setminus \setminus S_B$ is an I-bundle, and does not contain any guts. On the other hand, under stronger hypotheses about the exponents r_j, one can get non-trivial volume estimates from the guts of the other state surface S_A.

Theorem 9.7. *Let* $D = D_b$ *be a diagram of a hyperbolic link* K, *obtained as the closure of a positive braid* $b = \sigma_{i_1}^{r_1} \sigma_{i_2}^{r_2} \cdots \sigma_{i_k}^{r_k}$. *Suppose that* $r_j \geq 3$ *for all* $1 \leq j \leq k$; *in other words, each of the* k *twist regions in* D *contains at least* 3 *crossings. Then*

$$\frac{2v_8}{3} t(D) \leq \mathrm{vol}(S^3 \setminus K) < 10v_3(t(D) - 1),$$

where $v_3 = 1.0149\ldots$ *is the volume of a regular ideal tetrahedron and* $v_8 = 3.6638\ldots$ *is the volume of a regular ideal octahedron.*

Recall that $t(D)$ denotes the *twist number*: the number of twist regions in the diagram D. Observe that the multiplicative constants in the upper and lower bounds differ by a rather small factor of $4.155\ldots$.

The proof of Theorem 9.7 will require two lemmas.

Lemma 9.8. $D = D_b$ *be a diagram of a hyperbolic link* K, *obtained as the closure of the positive braid* $b = \sigma_{i_1}^{r_1} \sigma_{i_2}^{r_2} \cdots \sigma_{i_k}^{r_k}$, *where* $k \geq 2$. *Then*

(1) *If* K *is hyperbolic and* $r_j \geq 2$ *for all* j, *then* D *is a prime, A-adequate diagram.*
(2) *If* D *is a prime diagram and* $r_j \geq 6$, *for all* j, *then* K *is hyperbolic.*

Proof. First, suppose that $r_j \geq 2$ for all j. Since b is a positive braid, the A-resolution of every twist region is the long resolution (see Fig. 5.4 on p. 86). Thus every edge of $\mathbb{G}_A$ connects to a bigon on at least one end, and no edge of $\mathbb{G}_A$ is a loop. Thus D is A-adequate.

If K is hyperbolic, it must be prime and non-split. Thus, by Corollary 3.21 on p. 48, either D is prime or contains nugatory crossings. But a nugatory crossing in a braid diagram can only be created by stabilization, which would imply there is a term σ_i^1, contradicting the hypothesis that $r_j \geq 2$ for all j. This proves (1).

Statement (2) follows immediately from [38, Theorem 1.4], once one knows that D is twist-reduced. As we will not need conclusion (2) in the sequel (it was mainly included as a pleasant quasi-converse to (1)), we leave it to the reader to show that D is twist-reduced. □

Lemma 9.9. *Let $D = D_b$ be a diagram of a hyperbolic link K, obtained as the closure positive braid $b = \sigma_{i_1}^{r_1} \sigma_{i_2}^{r_2} \cdots \sigma_{i_k}^{r_k}$. Suppose that $r_j \geq 3$ for all j. Then D is A-adequate, and*

$$\chi(\mathbb{G}_A') = \chi(\mathbb{G}_A) \leq -2k/3 = -2t(D)/3 < 0.$$

Proof. The diagram D is A-adequate by Lemma 9.8. Since the A-resolution is the long resolution, every loop in $\mathbb{G}_A$ has length at least 3. Thus $\mathbb{G}_A = \mathbb{G}_A'$. It remains to count the vertices and edges of $\mathbb{G}_A$.

Recall that the edges of $\mathbb{G}_A$ are in one-to-one correspondence with the crossings in D_b; thus there are a total of $\sum r_j$ edges of $\mathbb{G}_A$. The vertices of $\mathbb{G}_A$ are in one-to-one correspondence with the state circles in the A-resolution of D_b. In a twist region with r_j crossings, there are $(r_j - 1)$ bigon circles in the long resolution; thus there are a total of $(\sum r_j) - k$ bigon state circles. It remains to count the non-bigon state circles of the A-resolution. We call these the *wandering* state circles, as they wander through multiple twist regions.

Consider the S^1-valued height function on $D(K)$ that arises from the braid position of the diagram. Relative to this height function, all segments of H_A are vertical, and connect two critical points of state circles. Thus the number of critical points on a state circle C equals the number of segments of H_A (equivalently, edges of G_A) met by C. To complete the proof of the lemma we need the following.

Claim. Every wandering state circle C has at least 6 critical points.

Proof of claim: Since C has the same number of minima as maxima, the total number of critical points on C must be even. Also, note that between critical points, C runs directly along one of the n strands of the braid. At a critical point, it crosses from the j-th to the $(j \pm 1)$-st strand.

Consider the number of distinct strands that C runs along. If C only runs along one strand of the braid, with no critical points, then that strand is a link component with no crossings: absurd. If C only runs along the i-th and $(i + 1)$-st strands of the braid, then it must have exactly 2 critical points, and is a bigon. This contradicts the hypothesis that C is wandering.

If C runs along four or more strands of the braid, then it needs at least 6 critical points (to get from the i-th to the $(i + 3)$-rd strand, and back), hence we are done. The remaining possibility is that C runs along exactly three strands of the braid. This means that C must have at least 4 critical points. If it has more than 4, then we are done.

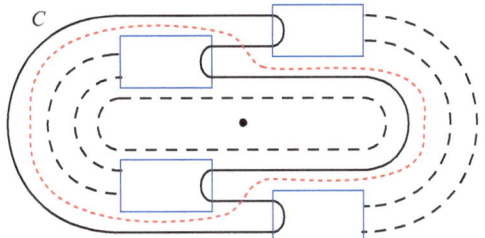

Fig. 9.2 A wandering state circle C with exactly 4 critical points. The *rectangular boxes* are twist regions. The *dashed arcs* are strands of the braid, which may run through other twist regions. The *red dotted* loop provides a contradiction to primeness

Suppose, for a contradiction, that C runs along exactly three strands and has exactly 4 critical points. Then, with some choice of orientation along C, it must run from the i-th to the $(i + 1)$-st strand at a maximum, then to the $(i + 2)$-nd at a minimum, then to the $(i + 1)$-st strand at a maximum, then finally back to the i-th at a minimum. In other words, C must look exactly like the state circle of Fig. 9.2. But the figure reveals an essential loop (dotted, red) meeting $D(K)$ twice, which contradicts primeness. Since D is prime by Lemma 9.8, this finishes the proof of the claim.

To continue with the proof of the lemma, observe that every twist region contains two critical points of wandering state circles; these are the ends of the long resolution in Fig. 5.4 on p. 86. On the other hand, by the claim, each wandering circle has at least 6 critical points. Thus there must be at least three times as many twist regions as wandering circles. We may now compute:

$$\chi(\mathbb{G}'_A) = \chi(\mathbb{G}_A) = \text{(bigon circles)} + \text{(wandering circles)} - \text{(crossings)}$$
$$\leq \left(\sum r_j - k\right) + (k/3) - \left(\sum r_j\right)$$
$$= -\frac{2k}{3} = -\frac{2t(D)}{3}.$$ $\square$

We can now complete the proof of Theorem 9.7.

Proof (of Theorem 9.7). The upper bound on volume is due to Agol and Thurston [58, Appendix], and holds for all diagrams.

To prove the lower bound on volume, recall that D must be prime by Lemma 9.8. By Lemma 9.9, we know that $\mathbb{G}_A = \mathbb{G}'_A$, hence $\mathbb{G}_A$ has no 2-edge loops. Thus Corollary 9.4 applies. Plugging the estimate

$$\chi_-(\mathbb{G}'_A) = -\chi(\mathbb{G}'_A) \geq 2t(D)/3$$

into Corollary 9.4 completes the proof. $\square$

Similar relations between volume and twist number are known for alternating links, links that admit diagrams with at least seven crossings in each twist region, and closed 3-braids [6, 32, 34]. To this list, we may add a result about the volumes of Montesinos links.

9.2   Volumes of Montesinos Links

In this section, we will prove Theorem 9.12, which estimates the volume of Montesinos links. We begin with a pair of lemmas. For the statement of the lemmas, recall Definition 8.3 on p. 122 and Definition 8.5 on p. 122.

Lemma 9.10. *Let $D(K)$ be a reduced, admissible Montesinos diagram with at least three positive tangles and at least three negative tangles. Let $\mathbb{G}'_A$ and $\mathbb{G}'_B$ be the reduced all-A and all-B graphs associated to D. Then*

$$-\chi(\mathbb{G}'_A) - \chi(\mathbb{G}'_B) = t(D) - Q_{1/2}(D),$$

where $Q_{1/2}(D)$ is the number of rational tangles in D whose slope has absolute value $|q| \in [1/2, 1)$.

Proof. The link diagram D can be used to construct a *Turaev surface* T: this is a closed, unknotted surface in S^3, onto which K has an alternating projection. The graphs $\mathbb{G}_A$ and $\mathbb{G}_B$ naturally embed in T as checkerboard graphs of the alternating projection, and are dual to one another. Furthermore, because D is constructed as a cyclic sum of alternating tangles, the Turaev surface is a torus. See [21, Sect. 4] for more details.

Recall that $\chi(\mathbb{G}'_A) = v_A - e'_A$, where v_A is the number of vertices and e_A is the number of edges, and similarly for $\chi(\mathbb{G}'_B)$. We can use the topology of T to get a handle on these quantities. Because $\mathbb{G}_A$ and $\mathbb{G}_B$ are dual, the number of vertices of $\mathbb{G}_B$ equals the number of regions in the complement of $\mathbb{G}_A$. Thus, since T is a torus, we have

$$v_A - e_A + v_B = \chi(T) = 0.$$

Now, consider the number of edges of $\mathbb{G}_A$ that are discarded when we pass to $\mathbb{G}'_A$. Because D has at least three positive tangles, Lemma 8.14 on p. 128 implies that edges can be lost in one of two ways:

(1) If r is an A-region with $c(r) > 1$ crossings, hence $c(r) > 1$ parallel edges in $\mathbb{G}_A$, then $c(r) - 1$ of these edges will be discarded as we pass to $\mathbb{G}'_A$. See Definition 5.16 and Fig. 5.4 on p. 86.

(2) If N_i is a negative tangle of slope $q_i \in (-1, -1/2]$, then one edge of $\mathbb{G}_A$ will be lost from the two-edge loop that spans N_i north to south. See Fig. 8.6 on p. 129.

The same principle holds for the B-graph $\mathbb{G}_B$, with B-regions replacing A-regions and positive tangles replacing negative ones.

Combining these facts gives

$$(e_A - e'_A) + (e_B - e'_B) = \sum_{\text{twist regions}} (c(r) - 1) \ + \#\{i : |q_i| \in [1/2, 1)\}$$

$$= c(D) - t(D) + Q_{1/2}(D).$$

Finally, since the edges of $\mathbb{G}_B$ are in one-to-one correspondence with the crossings of D,

$$
\begin{aligned}
-\chi(\mathbb{G}'_A) - \chi(\mathbb{G}'_B) &= e'_A + e'_B - v_A - v_B \\
&= (e'_A + e'_B - e_A - e_B) \quad + e_B \quad + (e_A - v_A - v_B) \\
&= -c(D) + t(D) - Q_{1/2} \quad + c(D) \quad + 0 \\
&= t(D) - Q_{1/2}(D).
\end{aligned}
$$

$\square$

Lemma 9.11. *Let $D(K)$ be a reduced, admissible Montesinos diagram with at least three positive tangles and at least three negative tangles. Then*

$$-\chi(\mathbb{G}'_A) - \chi(\mathbb{G}'_B) \geq \frac{t(D) - \#K}{2}.$$

where $\#K$ is the number of link components of K.

Proof. By Lemma 9.10, it will suffice to estimate the quantity $Q_{1/2}(D)$. Consider a rational tangle R_i whose slope satisfies $|q_i| \in [1/2, 1)$. Each such tangle contributes one unit to the count $Q_{1/2}(D)$. If $|q_i| > 1/2$, then the continued fraction expansion of q_i has at least two terms, hence R_i has at least two twist regions. Only one of those twist regions will be lost to the count $Q_{1/2}(D)$.

Alternately, suppose $q_i = \pm 1/2$. In this case, one strand of K in this tangle runs from the NW to the SW corner of the tangle, and another strand runs from the NE to the SE corner. In other words, the number of link components of K will remain unchanged if we replace R_i by a tangle of slope ∞. See Fig. 8.1 on p. 120.

Let n be the number of tangles of slope $\pm 1/2$ in the diagram D. If we replace each such tangle by one of slope ∞, the number $\#K$ of link components is unchanged. But after this replacement, there are n "breaks" in the diagram, hence K is a link of at least n components. This proves that $n \leq \#K$. In other words, there is a one-to-one mapping from tangles of slope $\pm 1/2$ to link components. We conclude that

$$Q_{1/2}(D) = \sum_{i:|q_i|>1/2} 1 + \sum_{i:|q_i|=1/2} 1$$

$$\leq \sum_{i:|q_i|>1/2} \frac{t(R_i)}{2} + \sum_{i:|q_i|=1/2} \frac{t(R_i)+1}{2}$$

$$\leq \frac{t(D) + \#K}{2},$$

and the result follows by Lemma 9.10.   □

Theorem 9.12. *Let $K \subset S^3$ be a Montesinos link with a reduced Montesinos diagram $D(K)$. Suppose that $D(K)$ contains at least three positive tangles and at least three negative tangles. Then K is a hyperbolic link, satisfying*

$$\frac{v_8}{4} \left(t(D) - \#K \right) \leq \mathrm{vol}(S^3 \setminus K) < 2v_8 \, t(D),$$

where $v_8 = 3.6638\ldots$ is the volume of a regular ideal octahedron and $\#K$ is the number of link components of K. The upper bound on volume is sharp.

We note that the upper bound on volume applies to all Montesinos links, without any restriction on the number of positive and negative tangles.

The lower bound on volume is proved using Lemma 9.11. In fact, using Lemma 9.10 instead of Lemma 9.11, one can obtain the sharper estimate

$$\mathrm{vol}(S^3 \setminus K) \geq \frac{v_8}{2} \left(t(D) - Q_{1/2}(D) \right),$$

where $Q_{1/2}(D)$ is the number of rational tangles of slope $|q_i| \in [1/2, 1)$.

Proof. Let $D(K)$ be a reduced Montesinos diagram that contains at least three positive tangles and at least three negative tangles. As we have observed following Definition 8.5 on p. 122, any reduced diagram can be made admissible by a sequence of flypes. Since flyping does not change the twist number of D, we may also assume that D is admissible. Thus Theorem 8.6 and Lemmas 9.10 and 9.11 all apply to $D(K)$.

The hyperbolicity of K follows from Bonahon and Siebenmann's enumeration of non-hyperbolic arborescent links [14]. See also Futer and Guéritaud [30, Theorem 1.5].

The lower bound on volume follows quickly by applying Theorem 9.1 to both the all-A and all-B state surfaces:

$$\mathrm{vol}(S^3 \setminus K) \geq v_8/2 \left(\chi_-\mathrm{guts}(S^3 \backslash\backslash S_A) + \chi_-\mathrm{guts}(S^3 \backslash\backslash S_B) \right)$$

$$= v_8/2 \left(\chi_-(\mathbb{G}'_A) + \chi_-(\mathbb{G}'_B) \right)$$

$$\geq v_8/4 \left(t(D) - \#K \right),$$

where the last two lines used Theorem 8.6 and Lemma 9.11.

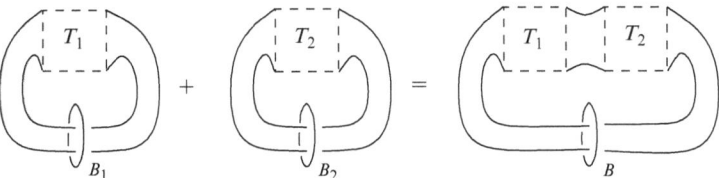

Fig. 9.3 A belted sum of tangles T_1 and T_2. The twice-punctured disks bounded by B_1 and B_2 are glued to form the twice-punctured disk bounded by the belt B

The upper bound on volume will follow from a standard Dehn filling argument. Add a link component B to K, which encircles the two eastern ends of some rational tangle T_i. Note that B can be moved by isotopy to lie between any pair of consecutive tangles. Thus the longitude of B forms (part of) the boundary of n different twice-punctured disks in $S^3 \setminus (K \cup B)$, with one disk between every pair of consecutive tangles.

The link $K \cup B$ is arborescent, hence also hyperbolic by [30]. Each twice-punctured disk bounded by B will be totally geodesic in this hyperbolic metric, by a theorem of Adams [1].

Let L_i be the link obtained by taking the numerator closure of tangle T_i, and adding an extra circle B_i about the eastern ends of the tangle. Then $(K \cup B)$ is a *belted sum* of the tangles $T_1, \ldots, T_n$: it is obtained by cutting each $S^3 \setminus L_i$ along the twice-punctured disk bounded by B_i, and gluing these manifolds cyclically along the twice-punctured disks. (See Fig. 9.3, and see Adams [1] for more information about belted sums.)

Since the numerator closure of each rational tangle T_i is a 2-bridge link, the link L_i is an augmented 2-bridge link. Thus each L_i is hyperbolic. Furthermore, if tangle T_i contains $t(T_i)$ twist regions, then L_i is the augmentation of a 2-bridge link with at most $t(T_i) + 1$ twist regions. Therefore, [45, Theorem B.3] implies that

$$\mathrm{vol}(S^3 \setminus L_i) < 2v_8 \, t(T_i).$$

When we perform the belted sum to obtain $K \cup B$, we cut and reglue along totally geodesic twice-punctured disks. Volume is additive under this operation [1]. This gives the estimate

$$\mathrm{vol}(S^3 \setminus (K \cup B)) < 2v_8 \, t(D).$$

Since volume goes down when we Dehn fill the meridian of B, the upper bound on $\mathrm{vol}(S^3 \setminus K)$ follows.

To prove the sharpness of the upper bound, consider the following sequence of examples. Let n be an even number, and let K_n be a Montesinos link with n rational tangles, where the slope of the j-th tangle is $(-1)^j / n$. This is a $(n, -n, \ldots, n, -n)$

pretzel link. Then every rational tangle has slope $\pm 1/n$, with alternating signs. The diagram D_n has exactly n twist regions, with exactly n crossings in each region.

Let J_n be a link obtained by adding a crossing circle about every twist region, as well as the belt component B of Fig. 9.3. Then, by the above discussion, J_n is a belted sum of n copies of the Borromean rings, hence

$$\mathrm{vol}(S^3 \setminus J_n) \; = \; 2v_8 \, n \; = \; 2v_8 \, t(D_n).$$

Furthermore, K_n can be recovered from J_n by $(\pm 1, n/2)$ Dehn filling on the crossing circles and meridional Dehn filling on the belt component B.

By [38, Theorem 3.8], there is an embedded horospherical neighborhood of the cusps of J_n, such that in each of the many 3-punctured spheres in $S^3 \setminus J_n$, the cusp neighborhoods of the 3 punctures are pairwise tangent. Then, by [38, Corollary 3.9 and Theorem 3.10], the Dehn filling curves have the following length on the horospherical tori:

- The meridian of the belt B has length $\ell(\mu) \geq n$,
- The $(\pm 1, n/2)$ curves on the crossing circles have length $\ell \geq \sqrt{n^2 + 1}$.

In particular, each filling curve has length at least n.

Now, we may use [32, Theorem 1.1] to bound the change in volume under Dehn filling. As a corollary of that theorem, it follows that when several cusps of a manifold M are filled along slopes of length at least $\ell_{\min} > 2\pi$, the additive change in volume satisfies

$$\Delta V \; \leq \; \frac{6\pi \cdot \mathrm{vol}(M)}{\ell_{\min}^2}.$$

(Deriving this corollary requires expanding the Taylor series for $(1 - x)^{3/2}$; see [32, Sect. 2.3].) In our setting, $\mathrm{vol}(S^3 \setminus J_n) = 2v_8 \, n$, and all the filling curves also have length at least n. Thus

$$2v_8 \, t(D_n) - \mathrm{vol}(S^3 \setminus K_n) \; = \; \Delta V \; \leq \; (6\pi \cdot 2v_8 \cdot n)/n^2,$$

which becomes arbitrarily small as $n \to \infty$. Thus the upper bound on volume is sharp. $\square$

Remark 9.13. The bounds on the change of volume under Dehn filling, obtained in [32], can be fruitfully combined with the results of this chapter. This combination results in relations between simple diagrammatic quantities (such as the twist number of an A-adequate knot) and the hyperbolic volume of 3-manifolds obtained by Dehn surgery on a that knot. For example, one may combine Theorem 9.12 with [32, Theorem 3.4] and obtain the following: Let $K \subset S^3$ be a Montesinos knot as in Theorem 9.12, and let N be a hyperbolic manifold obtained by p/q-Dehn surgery along K, where $|q| \geq 12$. Then

$$\frac{v_8}{4}\left(1 - \frac{127}{q^2}\right)^{3/2} (t(D) - 1) \; < \; \mathrm{vol}(N) \; < \; 2v_8 \, t(D).$$

9.3   Essential Surfaces and Colored Jones Polynomials

For a knot K let

$$J_K^n(t) = \alpha_n t^{m_n} + \beta_n t^{m_n - 1} + \ldots + \beta_n' t^{r_n + 1} + \alpha_n' t^{r_n},$$

denote its n-th *colored Jones polynomial*. One recently observed relation between the colored Jones polynomials and classical topology is the *slope conjecture* of Garoufalidis [42], which postulates that the sequence of degrees of the colored Jones polynomials detects certain boundary slopes of a knot K. This conjecture has been proved for several classes of knots [26, 36, 42], including a proof by the authors for the family of adequate knots [36]. See Theorem 1.6 in the Introduction for a precise statement.

In the same spirit, we can now show that certain coefficients of $J_K^n(t)$ measure how far the surface S_A is from being a fiber. We need the following lemma; a similar statement holds for B-adequate diagrams.

Lemma 9.14. *Let $D = D(K)$ be an A-adequate diagram with reduced all-A state graph $\mathbb{G}_A'$. Then for every $n > 1$*

(1) $|\alpha_n'| = 1$; *and*
(2) $|\beta_n'| = 1 - \chi(\mathbb{G}_A')$,

where as above α_n', β_n' are the last and next-to-last coefficients of $J_K^n(t)$.

Proof. Part (1) is proved in [60]; part (2) is proved in [23]. See [22] for an alternate proof of both results. □

Definition 9.15. With the setting and notation of Lemma 9.14, we define the *stable penultimate coefficient* of $J_K^n(t)$ to be $\beta_K' := |\beta_n'|$, for $n > 1$. For completeness we define the *stable last coefficient* to be $\alpha_K' := |\alpha_n'| = 1$.

We also define $\epsilon_K' = 1$ if $\beta_K' = 0$ and 0 otherwise.

Similarly, for a B-adequate knot K, we define the *stable second coefficient* of $J_K^n(t)$ to be $\beta_K := |\beta_n|$, for $n > 1$, and the *stable first coefficient* to be $\alpha_K := |\alpha_n| = 1$. We also define $\epsilon_K = 1$ if $\beta_K = 0$ and 0 otherwise.

The next result, which is a corollary of Theorem 5.11, shows that the stable coefficients β_K', β_K are exactly the obstructions to S_A or S_B being fibers. We only state the result for A-adequate links.

Corollary 9.16. *For an A-adequate link K, the following are equivalent:*

(1) $\beta_K' = 0$.
(2) *For every A-adequate diagram of $D(K)$, $S^3 \setminus K$ fibers over S^1 with fiber the corresponding state surface $S_A = S_A(D)$.*
(3) *For some A-adequate diagram $D(K)$, $M_A = S^3 \backslash\backslash S_A$ is an I-bundle over $S_A(D)$.*

Proof. By Lemma 9.14, $\beta_K' = 0$ precisely when $\mathbb{G}_A'$ is a tree, for every A-adequate diagram of K. Thus (1) $\Rightarrow$ (2) follows immediately from Theorem 5.11 on p. 82. The implication (2) $\Rightarrow$ (3) is trivial, by specializing to a particular A-adequate diagram. Finally, (3) $\Rightarrow$ (1) is again immediate from Theorem 5.11 and Lemma 9.14. $\square$

Remark 9.17. Given a knot K, the *Seifert genus* $g(K)$ is defined to be the smallest genus over all orientable surfaces spanned by K. Since a fiber realizes the genus of a knot [15], Corollary 9.16 implies that $g(K)$ can be read off from any A-adequate diagram of K when $\beta_K' = 0$. Since $\mathbb{G}_A$ is a spine for S_A (Lemma 2.2), in this case we have

$$g(K) = \frac{1 - \chi(\mathbb{G}_A)}{2}.$$

Note that $|\beta_K'| - 1 + \epsilon_K' = 0$ precisely when $|\beta_K'| \in \{0, 1\}$. By Corollary 9.16, having $\beta_K' = 0$ corresponds to S_A being a fiber. Our next result is that $|\beta_K'| = 1$ precisely when M_A is a book of I-bundles (hence, S_A is a fibroid) of a particular type.

Theorem 9.18. *Let K be an A-adequate link, and let β_K' be as in Definition 9.15. Then the following are equivalent:*

(1) $|\beta_K'| = 1$.
(2) *For every A-adequate diagram of K, the corresponding 3-manifold M_A is a book of I-bundles, with $\chi(M_A) = \chi(\mathbb{G}_A) - \chi(\mathbb{G}_A')$, and is not a trivial I-bundle over the state surface S_A.*
(3) *For some A-adequate diagram of K, the corresponding 3-manifold M_A is a book of I-bundles, with $\chi(M_A) = \chi(\mathbb{G}_A) - \chi(\mathbb{G}_A')$.*

Proof. For (1) $\Rightarrow$ (2), suppose that $|\beta_K'| = 1$, and let D be an A-adequate diagram. Then, by Theorem 5.14 and Lemma 9.14,

$$\chi_-(\text{guts}(M_A)) \; = \; \chi_-(\mathbb{G}_A') - \|E_c\| \; = \; 1 - |\beta_K'| - \|E_c\| \; \leq \; 0.$$

Since $\chi_-(\cdot) \geq 0$ by definition, it follows that $\chi(\text{guts}(M_A)) = 0$, hence $\text{guts}(M_A) = \emptyset$. In other words, if there are no guts, all of M_A is a book of I-bundles. But, by Corollary 9.16, M_A cannot be an I-bundle over S_A, because $|\beta_K'| \neq 0$.

(2) $\Rightarrow$ (3) is trivial.

For (3) $\Rightarrow$ (1), suppose that for some A-adequate diagram, M_A is a book of I-bundles, satisfying $\chi(M_A) = \chi(\mathbb{G}_A) - \chi(\mathbb{G}_A')$. By (5.1) on p. 85, this means $\chi(\mathbb{G}_A') = 0$. Thus, by Lemma 9.14, $|\beta_K'| = 1$. $\square$

One of the main results in this manuscript is the following theorem, which shows that β_K' monitors the topology of M_A quite effectively.

Theorem 9.19. *Let $D = D(K)$ be a prime A-adequate diagram of a link K with prime polyhedral decomposition of $M_A = S^3 \backslash\backslash S_A$ and let β_K' and and ϵ_K' be as in*

Definition 9.15. Then we have

$$\|E_c\| + \chi_-(\text{guts}(M_A)) = |\beta'_K| - 1 + \epsilon'_K,$$

where $\|E_c\|$ *is as in Definition 5.9, on p. 81.*

There is a similar statement for the stable second coefficient of $J^n_K(t)$ of B-adequate links. If $D = D(K)$ be a prime B-adequate diagram of a link K, then

$$\|E_c\| + \chi_-(\text{guts}(M_B)) = |\beta_K| - 1 + \epsilon_K,$$

where again $\|E_c\|$ is the smallest number of complex disks required to span the I-bundle of the upper polyhedron, as in Definition 5.9 on p. 81.

Proof. By Theorem 5.14, p. 84, we have

$$\chi_-(\text{guts}(M_A)) + \|E_c\| = \chi_-(\mathbb{G}'_A).$$

By Definition 1.5 on p. 6,

$$\chi_-(\mathbb{G}'_A) = -\chi(\mathbb{G}'_A) + \chi_+(\mathbb{G}'_A),$$

where $\chi_+(\mathbb{G}'_A) = 1$ if $\mathbb{G}'_A$ is tree and 0 otherwise. By Lemma 9.14(2)

$$|\beta'_K| - 1 = -\chi(\mathbb{G}'_A),$$

which implies that $|\beta'_K| = 0$ if and only if $\mathbb{G}'_A$ is a tree. This in turn implies that $|\beta'_K| = 0$ if and only if $\chi_+(\mathbb{G}'_A) = 1$. Combining all these we see that the quantity

$$\epsilon'_K := \chi_-(\text{guts}(M_A)) + \|E_c\| - |\beta'_K| + 1,$$

is equal to 1 if $|\beta'_K| = 0$ and 0 otherwise. This proves the equation in the statement of the theorem.   □

A simpler version of Theorem 9.19 is Theorem 9.20, which was stated in the introduction.

Theorem 9.20. *Suppose K is an A-adequate link whose stable penultimate coefficient is $\beta'_K \neq 0$. Then, for every A-adequate diagram $D(K)$,*

$$\chi_-(\text{guts}(M_A)) + \|E_c\| = |\beta'_K| - 1,$$

where $\|E_c\| \geq 0$ *is as in Definition 5.9. Furthermore, if D is prime and every 2-edge loop in $\mathbb{G}_A$ has edges belonging to the same twist region, then $\|E_c\| = 0$ and*

$$\chi_-(\text{guts}(M_A)) = |\beta'_K| - 1.$$

Proof. The first equation of the theorem follows immediately from Theorem 9.19, since $\epsilon'_K = 0$ when $\beta'_K \neq 0$. The second equation of the theorem follows by combining Corollary 5.19 with Lemma 9.14, since $\|E_c\| = 0$ when the edges of each 2-edge loop in $\mathbb{G}_A$ the edges belong in the same twist region of the diagram.   □

When $\|E_c\| = 0$, Theorem 9.19 provides particularly striking evidence that coefficients of the Jones polynomials measure something quite geometric: when $|\beta'_K|$ is large, the link complement $S^3 \setminus K$ contains essential surfaces that are correspondingly far from being fibroids. As a result, if K is hyperbolic, $S^3 \setminus K$ is forced to have large volume. As noted earlier, classes of links with $\|E_c\| = 0$ include alternating knots and Montesinos links. In this case we have the following.

Corollary 9.21. *Suppose K is a Montesinos link with a reduced admissible diagram $D(K)$ that contains at least three tangles of positive slope. Then*

$$\chi_-(\text{guts}(M_A)) = |\beta'_K| - 1.$$

Similarly, if $D(K)$ contains at least three tangles of negative slope, then

$$\chi_-(\text{guts}(M_B)) = |\beta_K| - 1.$$

Proof. Suppose that $D(K)$ has $r \geq 3$ tangles of positive slope. Then Theorem 8.6 on p. 123 implies that $\chi_-(\text{guts}(M_A)) = \chi_-(\mathbb{G}'_A)$. Furthermore, observe in Fig. 8.5 on p. 124 that the graph $\mathbb{G}_A$ contains at least one loop of length $r \geq 3$; this is the loop that spans every positive tangle west to east. All the edges of this loop are distinct in $\mathbb{G}'_A$. Thus $\mathbb{G}'_A$ contains at least one non-trivial loop, and is not a tree. Therefore, by Lemma 9.14 on p. 149,

$$\chi_-(\text{guts}(M_A)) = \chi_-(\mathbb{G}'_A) = -\chi(\mathbb{G}_A) = |\beta'_K| - 1.$$

The argument for three negative tangles is identical.   □

9.4 Hyperbolic Volume and Colored Jones Polynomials

If the volume conjecture is true, then for large n, it would imply a relation between the volume of a knot complement $S^3 \setminus K$ and coefficients of $J^n_K(t)$. For example, for $n \gg 0$ one would have $\text{vol}(S^3 \setminus K) < C\|J^n_K\|$, where $\|J^n_K\|$ denotes the L^1-norm of the coefficients of $J^n_K(t)$ and C is an appropriate constant. A series of articles written in recent years [23, 32–34] has established such relations for several classes of knots. In fact, in all the known cases the upper bounds on volume are paired with similar lower bounds. In several cases, our results here provide an intrinsic and satisfactory explanation for the existence of the lower bounds.

To illustrate this, let us look at the example of hyperbolic links K that have diagrams $D = D(K)$ that are positive closed braids, such that each twist region has at least seven crossings. As before, let β_K, β'_K denote the stable second and penultimate coefficients of $J^n_K(t)$ (Definition 9.15). Corollary 1.6 of [32] states that the quantity $|\beta_K| + |\beta'_K|$ provides two-sided bounds for the volume $\mathrm{vol}(S^3 \setminus K)$. As we saw in the discussion before Theorem 9.7, the graph $\mathbb{G}'_B$ is a tree, hence $\beta_K = 0$. Thus the two-sided bound on volume is in terms of $|\beta'_K|$ alone. However, since the argument of [32] is somewhat indirect (requiring twist number as an intermediate quantity), the upper and lower bounds differ by a factor of about 86.

Our results in this monograph (Corollary 5.19) reveal that the quantity $|\beta'_K| - 1$ realizes the guts of the incompressible surface S_A; hence, in the light of Theorem 9.1, we expect it to show up as a lower bound on the volume of $S^3 \setminus K$. In fact we can now show that $|\beta'_K| - 1$ gives two-sided bounds on the volume of positive braids that have only three crossings per twist region, rather than seven. Furthermore, because the argument using guts is more direct and intrinsic, the factor between the upper and lower bounds is now about 4.15.

Corollary 9.22. *Suppose that a hyperbolic link K is the closure of a positive braid $b = \sigma^{r_1}_{i_1} \sigma^{r_2}_{i_2} \cdots \sigma^{r_k}_{i_k}$, where $r_j \geq 3$ for all $1 \leq j \leq k$. Then*

$$v_8 \left(|\beta'_K| - 1 \right) \leq \mathrm{vol}(S^3 \setminus K) < 15 v_3 \left(|\beta'_K| - 1 \right) - 10 v_3,$$

where $v_3 = 1.0149\ldots$ is the volume of a regular ideal tetrahedron and $v_8 = 3.6638\ldots$ is the volume of a regular ideal octahedron.

Proof. By Lemma 9.9, the graph $\mathbb{G}_A$ has no 2-edge loops, and $\chi(\mathbb{G}_A) = \chi(\mathbb{G}'_A) < 0$. Thus, by Corollary 9.4,

$$\mathrm{vol}(S^3 \setminus K) \geq -v_8 \, \chi(\mathbb{G}'_A) = v_8(|\beta'_K| - 1).$$

For the upper bound on volume, we also use Lemma 9.9. The estimate of that lemma implies that

$$t(D) \leq -\tfrac{3}{2} \chi(\mathbb{G}'_A) = \tfrac{3}{2}|\beta'_K| - \tfrac{3}{2}.$$

Combined with Agol and D. Thurston's bound $\mathrm{vol}(S^3 \setminus K) < 10 v_3(t(D) - 1)$, this completes the proof.   □

A second result in this vein concerns Montesinos knots and links.

Corollary 9.23. *Let $K \subset S^3$ be a Montesinos link with a reduced Montesinos diagram $D(K)$. Suppose that $D(K)$ contains at least three positive tangles and at least three negative tangles. Then K is a hyperbolic link, satisfying*

$$v_8 \left(\max\{|\beta_K|, |\beta'_K|\} - 1 \right) \leq \mathrm{vol}(S^3 \setminus K) < 4 v_8 \left(|\beta_K| + |\beta'_K| - 2 \right) + 2 v_8 \, (\#K),$$

where $\#K$ is the number of link components of K.

We remark that the number of link components $\#K$ is recoverable from the Jones polynomial evaluated at 1: $J_K(1) = (-2)^{\#K-1}$. See [49].

Proof. The lower bound on volume is Theorem 9.1 combined with Corollary 9.21. For the upper bound on volume, we combine the upper bound of Theorem 9.12 with the estimate of Lemma 9.11:

$$
\begin{aligned}
\mathrm{vol}(S^3 \setminus K) &< 2v_8\, t(D) \\
&\leq 2v_8 \left(-2\chi(\mathbb{G}'_A) - 2\chi(\mathbb{G}'_A) + \#K \right) \\
&= 4v_8 \left(|\beta_K| + |\beta'_K| - 2 \right) + 2v_8\, (\#K).
\end{aligned}
$$

$\square$

Chapter 10
Discussion and Questions

In this final chapter, we state some questions that arose from this work and speculate about future directions related to this project. In Sect. 10.1, we discuss modifications of the diagram D that preserve A-adequacy. In Sect. 10.2, we speculate about using normal surface theory in our polyhedral decomposition of M_A to attack various open problems, for example the Cabling Conjecture and the determination of hyperbolic A-adequate knots. In Sect. 10.3, we discuss extending the results of this monograph to states other than the all-A (or all-B) state. Finally, in Sect. 10.4, we discuss a coarse form of the hyperbolic volume conjecture.

10.1 Efficient Diagrams

To motivate our discussion of diagrammatic moves, recall the well-known *Tait conjectures* for alternating links:

(1) Any two reduced alternating projections of the same link have the same number of crossings.
(2) A reduced alternating diagram of a link has the least number of crossings among all the projections of the link.
(3) Given two reduced, prime alternating diagrams D and D' of the same link, it is possible to transform D to D' by a finite sequence of *flypes*.

Statements (1) and (2) where proved by Kauffman [55] and Murasugi [74] using properties of the Jones polynomial. A shorter proof along similar lines was given by Turaev [95]. Statement (3), which is known as the "flyping conjecture" was proven by Menasco and Thistlethwaite [67]. Note that the Jones polynomial is also used in that proof.

One can ask to what extend the statements above can be generalized to semi-adequate links. It is easy to see that statements (1) and (2) are not true in this case: For instance, the two diagrams in Example 5.3 on p. 74 are both A-adequate, but have different numbers of crossings. Nonetheless, some information is known about

D. Futer et al., *Guts of Surfaces and the Colored Jones Polynomial*, Lecture Notes in Mathematics 2069, DOI 10.1007/978-3-642-33302-6_10,
© Springer-Verlag Berlin Heidelberg 2013

crossing numbers of semi-adequate diagrams: Stoimenow showed that the number of crossings of any semi-adequate projection of a link is bounded above by a link invariant that is expressed in terms of the 2-variable Kauffman polynomial and the maximal Euler characteristic of the link. As a result, he concluded that each semi-adequate link has only finitely many semi-adequate reduced diagrams [91, Theorem 1.1]. In view of his work, it seems natural to ask for an analogue of the flyping conjecture in the setting of semi-adequate links.

Problem 10.1. Find a set of diagrammatic moves that preserve A-adequacy and that suffice to pass between any pair of reduced, A-adequate diagrams of a link K.

A solution to Problem 10.1 would help to clarify to what extent the various quantities introduced in this monograph actually depend on the choice of A-adequate diagram $D(K)$. Recall the prime polyhedral decomposition of $M_A = S^3 \backslash\backslash S_A$ introduced above, and let β'_K and ϵ_K be as in Definition 9.15 on p. 149. Since $|\beta'_K| - 1 + \epsilon'_K$ is an invariant of K, Theorem 9.19 implies that the quantity $||E_c|| + \chi_-(\text{guts}(M_A))$ is also an invariant of K. As noted earlier, $||E_c||$ and $\chi_-(\text{guts}(M_A))$ are not, in general, invariants of K: they depend on the A-adequate diagram used. For instance, in Example 5.3 on p. 74, we show that by modifying the diagram of a particular link, we can eliminate the quantity $||E_c||$. This example, along with the family of Montesinos links (see Theorem 8.6 on p. 123), prompts the following question.

Question 10.2. Let K be a non-split, prime A-adequate link. Is there an A-adequate diagram $D(K)$, such that if we consider the corresponding prime polyhedral decomposition of $M_A(D) = S^3 \backslash\backslash S_A(D)$, we will have $||E_c|| = 0$? This would imply that
$$\chi_-(\text{guts}(M_A)) = \chi_-(\mathbb{G}'_A) = |\beta'_K| - 1 + \epsilon'_K.$$

Among the more accessible special cases of Question 10.2 is the following.

Question 10.3. Does Theorem 8.6 generalize to *all* Montesinos links? That is: can we remove the hypothesis that a reduced diagram $D(K)$ must contain at least three tangles of positive slope?

Note that if $D(K)$ has no positive tangles, then it is alternating, hence the conclusion of Theorem 8.6 is known by [58]. If $D(K)$ has one positive tangle, then it is not A-adequate by Lemma 8.4. Thus Question 10.3 is open only in the case where $D(K)$ contains exactly two tangles of positive slope.

Another tractable special case of Question 10.2 is the following.

Question 10.4. Let K be an A-adequate link that can be depicted by a diagram $D(K)$, obtained by Conway summation of alternating tangles. Each such link admits a Turaev surface of genus one [21]. Does there exist a (possibly different) diagram of K, for which $||E_c|| = 0$?

Prior to this manuscript, there have been only a few cases in which the guts of essential surfaces have been explicitly understood and calculated for an infinite family of 3-manifolds [3, 57, 58]. Affirmative answers to Questions 10.2, 10.3, and 10.4 would add to the list of these results, and could have further applications. In particular, combined with Theorem 9.1, they would lead to new relations between quantum invariants and hyperbolic volume.

Next, recall from the end of Sect. 5.5, that given an A-adequate diagram $D := D(K)$, we denote by D^n the n-cabling of D with the blackboard framing. If D is A-adequate then D^n is A-adequate for all $n \in \mathbb{N}$. Furthermore, we have

$$\chi(\mathbb{G}'_A(D^n)) = \chi(\mathbb{G}'_A(D)),$$

for all $n \geq 1$. In other words, the quantity $\chi(\mathbb{G}'_A)$ remains invariant under cabling [60, Chap. 5]. Recall, from Corollary 5.20 on p. 88, that the quantity $\chi_-\text{guts}(S^3 \backslash \backslash S^n_A) + ||E_c(D^n)||$ is also invariant under planar cabling. This prompts the following question.

Question 10.5. Let $D := D(K)$ be a prime, A-adequate diagram, of a link K. For $n \geq 1$, let D^n denote the n-cabling of D using the blackboard framing. Is it true that $||E_c(D^n)|| = ||E_c(D)||$, hence $\chi_-(\text{guts}(S^3 \backslash \backslash S^n_A)) = \chi_-(\text{guts}(S^3 \backslash \backslash S_A))$, for every n as above?

We note that an affirmative answer to Question 10.5 would provide an intrinsic explanation for the fact that the coefficient β'_n of the colored Jones polynomials stabilizes.

10.2 Control Over Surfaces

In Chaps. 4 and 5, we controlled pieces of the characteristic submanifold of M_A by putting them in normal form with respect to the polyhedral decomposition constructed in Chap. 3. The powerful tools of normal surface theory have been used (sometimes in disguise) to obtain a number of results about alternating knots and links: see, for example, [58, 59, 65, 66]. It seems natural to ask what other results in this vein can be proved for A-adequate knots and links.

One sample open problem that should be accessible using these methods is the following the following problem posed by Ozawa [76].

Problem 10.6. Prove that an A-adequate knot is prime if and only if *every* A-adequate diagram without nugatory crossings is prime.

Recall that one direction of the problem is Corollary 3.21 on p. 48: if K is prime and $D(K)$ has no nugatory crossings, then D must be prime. To attack the converse direction of the problem, one might try showing that if K is not prime, then an A-adequate diagram $D(K)$ cannot be prime.

Suppose that K is not prime, and $\Sigma \subset S^3 \setminus K$ is an essential, meridional annulus in the prime decomposition. Then, since S_A is also an essential surface, Σ can be moved by isotopy into a position where it intersects S_A in a collection of essential arcs. Thus, after Σ is cut along these arcs, it must intersect $M_A = S^3 \backslash \backslash S_A$ in a disjoint union of EPDs. Now, all the machinery of Chap. 4 can be used to analyze these EPDs, with the aim of proving that D must not be prime.

The same ideas can be used to attack other problems that depend on an understanding of "small" surfaces in the link complement. For example, if $\Sigma \subset S^3 \setminus K$ is an essential torus, then $\Sigma \cap S_A$ must consist of simple closed curves that are essential on both surfaces. Cutting Σ along these curves, we conclude that $\Sigma \cap M_A$ is a union of annuli, which are contained in the maximal I-bundle of M_A. Thus once again, the machinery of Chap. 4 can be brought to bear: by Lemma 4.6, each annulus intersects the polyhedra in normal squares, and so on. This leads to the following question.

Problem 10.7. Give characterization of hyperbolic A-adequate links in terms of their A-adequate diagrams.

We are aware of only three families of A-adequate diagrams that depict non-hyperbolic links. First, the standard diagram of a (p, q)-torus link (where $p > 0$ and $q < 0$) is a negative braid, hence A-adequate by the discussion following Definition 9.6 on p. 141. Second, by Corollary 3.21 on p. 48, a non-prime A-adequate diagram (without nugatory crossings) must depict a composite link. Third, a planar cable of (some of the components of) a link K in an A-adequate diagram D also produces an A-adequate diagram D^n, but clearly is not hyperbolic. Thus the following naïve question has a chance of a positive answer:

Question 10.8. Suppose $D(K)$ is a prime A-adequate diagram that is not a planar cable and not the standard diagram of a (p, q)-torus link. Is K necessarily hyperbolic?

A related open problem is the celebrated *Cabling Conjecture*, which implies that a hyperbolic knot K does not have any reducible Dehn surgeries. While the conjecture has been proved for large classes of knots [27, 62, 66, 88], including all non-hyperbolic knots, it is still a major open problem. Note that if a Dehn filling of a knot K does contain an essential 2-sphere, then $S^3 \setminus K$ must contain an essential planar surface Σ, whose boundary is the slope along which we perform the Dehn filling. The Cabling Conjecture asserts that K must be a cable knot and Σ is the *cabling annulus*. Given existing work [70, 88], an equivalent formulation is that hyperbolic knots do not have any reducible surgeries.

If K is an A-adequate knot, then our results here provide a nice ideal polyhedral decomposition of associated 3-manifold M_A. It would be interesting to attempt to analyze essential planar surfaces in $S^3 \setminus K$ by putting them in normal form with respect to this decomposition, to attack the following problem.

Problem 10.9. If Σ is an essential planar surface in the complement of an A-adequate knot K, show that either $\partial \Sigma$ consists of meridians of K, or Σ is a cabling annulus. That is, prove the Cabling Conjecture for A-adequate knots.

Recall that the class of A-adequate knots is very large; see Sect. 1.3 on p. 7. Therefore, the resolution of Problem 10.9 would be a major step toward a proof of the Cabling Conjecture.

10.3   Other States

As we mentioned in Chap. 2, one may associate many states to a link diagram. Any choice of state σ defines a state graph $\mathbb{G}_\sigma$ and a state surface S_σ (see also [36]). A natural and interesting question is: to what extent do the methods and results of this manuscript generalize to states other than the all-A and the all-B state? For example, one can ask the following question.

Question 10.10. Does every knot K admit a diagram $D(K)$ and a state σ so that S_σ is essential in $S^3 \setminus K$?

As we have seen in Sects. 2.4, 3.4, 4.5, and 5.6, all of our structural results about the polyhedral decomposition generalize to state surfaces of σ-homogeneous, σ-adequate states. In particular, the state surface S_σ of such a state must be essential, recovering Ozawa's Theorem 3.25. In [21], Dasbach, Futer, Kalfagianni, Lin, and Stoltzfus show that for any diagram $D(K)$, the entire Jones polynomial $J_K(t)$ can be computed from the Bollobás–Riordan polynomial [12, 13] of the *ribbon graph* associated to the all-A graph $\mathbb{G}_A$ or the all-B graph $\mathbb{G}_B$. It is natural to ask whether these results extend to other states.

Question 10.11. Let $D(K)$ be a link diagram that is σ-adequate and σ-homogeneous. Does the Bollobás–Riordan polynomial of the graph $\mathbb{G}_\sigma$ associated to σ carry all of the information in the Jones polynomial of K? How do these polynomials relate to the topology of the state surface S_σ?

10.4   A Coarse Volume Conjecture

Our results here, as well as several recent articles [23, 32–34], have established two-sided bounds on the hyperbolic volume of a link complement in terms of coefficients of the Jones and colored Jones polynomials. These results motivate the following question.

Definition 10.12. Let $f, g : Z \to \mathbb{R}_+$ be functions from some (infinite) set Z to the non-negative reals. We say that f and g are *coarsely related* if there exist universal constants $C_1 \geq 1$ and $C_2 \geq 0$ such that

$$C_1^{-1} f(x) - C_2 \leq g(x) \leq C_1 f(x) + C_2 \quad \forall x \in Z.$$

This notion is central in coarse geometry. For example, a function $\varphi : X \to Y$ between two metric spaces is a *quasi-isometric embedding* if $d_X(x, x')$ is coarsely related to $d_Y(\varphi(x), \varphi(x'))$. Here, $Z = X \times X$.

Question 10.13 (Coarse Volume Conjecture).   Does there exist a function $B(K)$ of the coefficients of the colored Jones polynomials of a knot K, such that for hyperbolic knots, $B(K)$ is coarsely related to hyperbolic volume?

Here, we are thinking of both vol $: Z \to \mathbb{R}_+$ and $B : Z \to \mathbb{R}_+$ as functions on the set Z of hyperbolic knots.

Work of Garoufalidis and Le [43] implies that for a given link K, the sequence $\{J_K^n(t)|n \in \mathbb{N}\}$ is determined by finitely many values of n. This implies that the coefficients satisfy linear recursive relations with constant coefficients [41]. For A-adequate links, the recursive relations between coefficients of $J_K^n(t)$ manifest themselves in the stabilization properties discussed in Lemma 9.14 on p. 149, and Definition 9.15 on p. 149. Lemma 9.14 is not true for arbitrary knots. However, numerical evidence and calculations (by Armond, Dasbach, Garoufalidis, van der Veen, Zagier, etc.) prompt the question of whether the first and last two coefficients of $J_K^n(t)$ "eventually" become periodic.

Question 10.14. Given a knot K, do there exist a "stable" integer $N = N(K) > 0$ and a "period" $p = p(K) > 0$, depending on K, such that for all $m \geq N$ where $m - N$ is a multiple of p,

$$|\alpha_m| = |\alpha_N|, \quad |\beta_m| = |\beta_N|, \quad |\beta'_m| = |\beta'_N|, \quad |\alpha'_m| = |\alpha'_N| ?$$

As discussed above, for knots that are both A and B-adequate, any integer $N \geq 2$ is "stable" with period $p = 1$. Examples show that in general, we cannot hope that $p = 1$ for arbitrary knots. For example, [9, Proposition 6.1] states that for several families of torus knots we have $p = 2$. In general, if the answer to Question 10.14 is *yes*, then if we take N to be the smallest "stable" integer then we may consider the $4p$ values

$$|\alpha_m|, \quad |\beta_m|, \quad |\beta'_m|, \quad |\alpha'_m|, \quad \text{for} \quad N \leq m \leq N + p - 1. \qquad (10.1)$$

The results [23, 32–34], as well as Corollary 9.23 in Chap. 9, prompt the question of whether this family of coefficients of $J_K^n(t)$ determines the volume of K up to a bounded constant.

Question 10.15. Suppose the answer to Question 10.14 is *yes*, and the stable values $|\alpha_m| |\beta_m|, |\beta'_m|, |\alpha'_m|$ of (10.1) are well-defined. Is there a function $B(K)$ of these stable coefficients that is coarsely related to the hyperbolic volume $\mathrm{vol}(S^3 \setminus K)$?

Remark 10.16. If K is an alternating knot then β_K, β'_K are equal to the second and penultimate coefficient of the ordinary Jones polynomial $J_K(t)$, respectively. Since the quantity $|\beta_K| + |\beta'_K|$ provides two sided bounds on the volume of hyperbolic alternating links one may wonder whether there is a function of the second and the penultimate coefficient of $J_K(t)$ that controls the volume of all hyperbolic knots K. In [34, Theorem 6.8], we show that is not the case. That is: there is no single function of the second and the penultimate coefficient of the Jones polynomial that can control the volume of all hyperbolic knots.

Finally, we note that the quantity on the right-hand side of the equation in the statement of Theorem 9.19 can be rewritten in the form $|\beta'_K| - |\alpha'_K| + \epsilon'_K$. In the view of this observation, it is tempting to ask whether analogues of Theorem 9.19 on p. 150 hold for *all* knots.

Question 10.17. Given a knot K for which the stable coefficients of Question 10.14 exist, is there an essential spanning surface S with boundary K such that the stable coefficients of (10.1) capture the topology of $S^3 \backslash\backslash S$ in the sense of Theorem 9.19, Corollary 9.16, and Theorem 9.18?

References

1. Adams, C.C.: Thrice-punctured spheres in hyperbolic 3-manifolds. Trans. Am. Math. Soc. **287**(2), 645–656 (1985)
2. Adams, C.C.: Noncompact Fuchsian and quasi-Fuchsian surfaces in hyperbolic 3-manifolds. Algebr. Geomet. Topology **7**, 565–582 (2007)
3. Agol, I.: Lower bounds on volumes of hyperbolic Haken 3-manifolds. arXiv:math/9906182
4. Agol, I.: The virtual Haken conjecture. arXiv:1204.2810. With an appendix by Ian Agol, Daniel Groves, and Jason Manning
5. Agol, I.: Criteria for virtual fibering. J. Topology **1**(2), 269–284 (2008). doi:10.1112/jtopol/jtn003
6. Agol, I., Storm, P.A., Thurston, W.P.: Lower bounds on volumes of hyperbolic Haken 3-manifolds. J. Am. Math. Soc. **20**(4), 1053–1077 (2007). With an appendix by Nathan Dunfield
7. Andreev, E.M.: Convex polyhedra in Lobačevskiĭ spaces. Mat. Sb. (N.S.) **81**(123), 445–478 (1970)
8. Andreev, E.M.: Convex polyhedra of finite volume in Lobačevskiĭ space. Mat. Sb. (N.S.) **83**(125), 256–260 (1970)
9. Armond, C., Dasbach, O.T.: Rogers–Ramanujan type identities and the head and tail of the colored Jones polynomial. arXiv:1106.3948
10. Atiyah, M.: The Geometry and Physics of Knots. Lezioni Lincee [Lincei Lectures]. Cambridge University Press, Cambridge (1990). doi:10.1017/CBO9780511623868
11. Atkinson, C.K.: Two-sided combinatorial volume bounds for non-obtuse hyperbolic polyhedra. Geometriae Dedicata **153**(1), 177–211 (2011)
12. Bollobás, B., Riordan, O.: A polynomial invariant of graphs on orientable surfaces. Proc. Lond. Math. Soc. (3) **83**(3), 513–531 (2001). doi:10.1112/plms/83.3.513
13. Bollobás, B., Riordan, O.: A polynomial of graphs on surfaces. Math. Ann. **323**(1), 81–96 (2002). doi:10.1007/s002080100297
14. Bonahon, F., Siebenmann, L.: New Geometric Splittings of Classical Knots, and the Classification and Symmetries of Arborescent Knots. Geometry and Topology Monographs (to appear). http://www-bcf.usc.edu/~fbonahon/Research/Preprints/Preprints.html
15. Burde, G., Zieschang, H.: Knots. de Gruyter Studies in Mathematics, vol. 5, 2nd edn. Walter de Gruyter, Berlin (2003)
16. Calegari, D., Freedman, M.H., Walker, K.: Positivity of the universal pairing in 3 dimensions. J. Am. Math. Soc. **23**(1), 107–188 (2010). doi:10.1090/S0894-0347-09-00642-0
17. Cha, J.C., Livingston, C.: Knotinfo: Table of knot invariants (2012) http://www.indiana.edu/~knotinfo
18. Champanerkar, A., Kofman, I., Patterson, E.: The next simplest hyperbolic knots. J. Knot Theor. Ramifications **13**(7), 965–987 (2004)

D. Futer et al., *Guts of Surfaces and the Colored Jones Polynomial*, Lecture Notes in Mathematics 2069, DOI 10.1007/978-3-642-33302-6, © Springer-Verlag Berlin Heidelberg 2013

19. Cromwell, P.R.: Homogeneous links. J. Lond. Math. Soc. (2) **39**(3), 535–552 (1989). doi:10.1112/jlms/s2-39.3.535

20. Culler, M., Shalen, P.B.: Volumes of hyperbolic Haken manifolds. I. Invent. Math. **118**(2), 285–329 (1994). doi:10.1007/BF01231535

21. Dasbach, O.T., Futer, D., Kalfagianni, E., Lin, X.S., Stoltzfus, N.W.: The Jones polynomial and graphs on surfaces. J. Combin. Theor. Ser. B **98**(2), 384–399 (2008)

22. Dasbach, O.T., Futer, D., Kalfagianni, E., Lin, X.S., Stoltzfus, N.W.: Alternating sum formulae for the determinant and other link invariants. J. Knot Theor. Ramifications **19**(6), 765–782 (2010)

23. Dasbach, O.T., Lin, X.S.: On the head and the tail of the colored Jones polynomial. Compos. Math. **142**(5), 1332–1342 (2006)

24. Dasbach, O.T., Lin, X.S.: A volume-ish theorem for the Jones polynomial of alternating knots. Pac. J. Math. **231**(2), 279–291 (2007)

25. Dimofte, T., Gukov, S.: Quantum field theory and the volume conjecture. Contemp. Math. **541**, 41–68 (2011)

26. Dunfield, N.M., Garoufalidis, S.: Incompressibility criteria for spun-normal surfaces. Trans. Am. Math. Soc. **364**(11), 6109–6137 (2012). doi:10.1090/S0002-9947-2012-05663-7

27. Eudave Muñoz, M.: Band sums of links which yield composite links. The cabling conjecture for strongly invertible knots. Trans. Am. Math. Soc. **330**(2), 463–501 (1992). doi:10.2307/2153918

28. Freyd, P., Yetter, D.N., Hoste, J., Lickorish, W.B.R., Millett, K.C., Ocneanu, A.: A new polynomial invariant of knots and links. Bull. Am. Math. Soc. (N.S.) **12**(2), 239–246 (1985). doi:10.1090/S0273-0979-1985-15361-3

29. Futer, D.: Fiber detection for state surfaces. arXiv:1201.1643 (2012)

30. Futer, D., Guéritaud, F.: Angled decompositions of arborescent link complements. Proc. Lond. Math. Soc. (3) **98**(2), 325–364 (2009). doi:10.1112/plms/pdn033

31. Futer, D., Kalfagianni, E., Purcell, J.S.: Quasifuchsian state surfaces. ArXiv:1209.5719 (2012)

32. Futer, D., Kalfagianni, E., Purcell, J.S.: Dehn filling, volume, and the Jones polynomial. J. Differ. Geom. **78**(3), 429–464 (2008)

33. Futer, D., Kalfagianni, E., Purcell, J.S.: Symmetric links and Conway sums: volume and Jones polynomial. Math. Res. Lett. **16**(2), 233–253 (2009)

34. Futer, D., Kalfagianni, E., Purcell, J.S.: Cusp areas of Farey manifolds and applications to knot theory. Int. Math. Res. Not. IMRN **2010**(23), 4434–4497 (2010)

35. Futer, D., Kalfagianni, E., Purcell, J.S.: On diagrammatic bounds of knot volumes and spectral invariants. Geometriae Dedicata **147**, 115–130 (2010). doi:10.1007/s10711-009-9442-6

36. Futer, D., Kalfagianni, E., Purcell, J.S.: Slopes and colored Jones polynomials of adequate knots. Proc. Am. Math. Soc. **139**, 1889–1896 (2011)

37. Futer, D., Kalfagianni, E., Purcell, J.S.: Jones polynomials, volume, and essential knot surfaces: a survey. In: Proceedings of Knots in Poland III. Banach Center Publications (to appear)

38. Futer, D., Purcell, J.S.: Links with no exceptional surgeries. Comment. Math. Helv. **82**(3), 629–664 (2007). doi:10.4171/CMH/105

39. Gabai, D.: The Murasugi sum is a natural geometric operation. In: Low-Dimensional Topology (San Francisco, CA, 1981). Contemporary Mathematics, vol. 20, pp. 131–143. American Mathematical Society, Providence (1983)

40. Gabai, D.: Detecting fibred links in S^3. Comment. Math. Helv. **61**(4), 519–555 (1986). doi:10.1007/BF02621931

41. Garoufalidis, S.: The degree of a q-holonomic sequence is a quadratic quasi-polynomial. Electron. J. Combin. **18**(2), Paper 4, 23 (2011)

42. Garoufalidis, S.: The Jones slopes of a knot. Quant. Topology **2**(1), 43–69 (2011). doi:10.4171/QT/13

43. Garoufalidis, S., Lê, T.T.Q.: The colored Jones function is q-holonomic. Geom. Topology **9**, 1253–1293 (electronic) (2005). doi:10.2140/gt.2005.9.1253

44. Gromov, M.: Volume and bounded cohomology. Inst. Hautes Études Sci. Publ. Math. (56), 5–99 (1982/1983)

45. Guéritaud, F., Futer, D. (appendix): On canonical triangulations of once-punctured torus bundles and two-bridge link complements. Geom. Topology **10**, 1239–1284 (2006)
46. Hoste, J., Thistlethwaite, M.B.: Knotscape (2012) http://www.math.utk.edu/~morwen
47. Jaco, W.H., Shalen, P.B.: Seifert fibered spaces in 3-manifolds. Mem. Am. Math. Soc. **21**(220), viii+192 (1979)
48. Johannson, K.: Homotopy Equivalences of 3-Manifolds with Boundaries. Lecture Notes in Mathematics, vol. 761. Springer, Berlin (1979)
49. Jones, V.F.R.: A polynomial invariant for knots via von Neumann algebras. Bull. Am. Math. Soc. (N.S.) **12**(1), 103–111 (1985). doi:10.1090/S0273-0979-1985-15304-2
50. Jones, V.F.R.: Hecke algebra representations of braid groups and link polynomials. Ann. Math. (2) **126**(2), 335–388 (1987)
51. Jørgensen, T.: Compact 3-manifolds of constant negative curvature fibering over the circle. Ann. Math. (2) **106**(1), 61–72 (1977)
52. Kashaev, R.M.: Quantum dilogarithm as a $6j$-symbol. Mod. Phys. Lett. A **9**(40), 3757–3768 (1994). doi:10.1142/S0217732394003610
53. Kashaev, R.M.: A link invariant from quantum dilogarithm. Mod. Phys. Lett. A **10**(19), 1409–1418 (1995). doi:10.1142/S0217732395001526
54. Kashaev, R.M.: The hyperbolic volume of knots from the quantum dilogarithm. Lett. Math. Phys. **39**(3), 269–275 (1997)
55. Kauffman, L.H.: State models and the Jones polynomial. Topology **26**(3), 395–407 (1987). doi:10.1016/0040-9383(87)90009-7
56. Kauffman, L.H.: An invariant of regular isotopy. Trans. Am. Math. Soc. **318**(2), 417–471 (1990). doi:10.2307/2001315
57. Kuessner, T.: Guts of surfaces in punctured-torus bundles. Arch. Math. (Basel) **86**(2), 176–184 (2006). doi:10.1007/s00013-005-1097-4
58. Lackenby, M.: The volume of hyperbolic alternating link complements. Proc. Lond. Math. Soc. (3) **88**(1), 204–224 (2004). With an appendix by Ian Agol and Dylan Thurston
59. Lackenby, M.: Classification of alternating knots with tunnel number one. Comm. Anal. Geom. **13**(1), 151–185 (2005)
60. Lickorish, W.B.R.: An Introduction to Knot Theory. Graduate Texts in Mathematics, vol. 175. Springer, New York (1997)
61. Lickorish, W.B.R., Thistlethwaite, M.B.: Some links with nontrivial polynomials and their crossing-numbers. Comment. Math. Helv. **63**(4), 527–539 (1988)
62. Luft, E., Zhang, X.: Symmetric knots and the cabling conjecture. Math. Ann. **298**(3), 489–496 (1994). doi:10.1007/BF01459747
63. Manchón, P.M.G.: Extreme coefficients of Jones polynomials and graph theory. J. Knot Theor. Ramifications **13**(2), 277–295 (2004). doi:10.1142/S0218216504003135
64. Menasco, W.W.: Polyhedra representation of link complements. In: Low-Dimensional Topology (San Francisco, CA, 1981). Contemporary Mathematics, vol. 20, pp. 305–325. American Mathematical Society, Providence (1983)
65. Menasco, W.W.: Closed incompressible surfaces in alternating knot and link complements. Topology **23**(1), 37–44 (1984). doi:10.1016/0040-9383(84)90023-5
66. Menasco, W.W., Thistlethwaite, M.B.: Surfaces with boundary in alternating knot exteriors. J. Reine Angew. Math. **426**, 47–65 (1992)
67. Menasco, W.W., Thistlethwaite, M.B.: The classification of alternating links. Ann. Math. (2) **138**(1), 113–171 (1993). doi:10.2307/2946636
68. Miyamoto, Y.: Volumes of hyperbolic manifolds with geodesic boundary. Topology **33**(4), 613–629 (1994). doi:10.1016/0040-9383(94)90001-9
69. Morgan, J., Tian, G.: Ricci Flow and the Poincaré Conjecture. Clay Mathematics Monographs, vol. 3. American Mathematical Society, Providence (2007)
70. Moser, L.: Elementary surgery along a torus knot. Pac. J. Math. **38**, 737–745 (1971)
71. Mostow, G.D.: Quasi-conformal mappings in n-space and the rigidity of hyperbolic space forms. Inst. Hautes Études Sci. Publ. Math. (34), 53–104 (1968)

72. Murakami, H.: An introduction to the volume conjecture. In: Interactions Between Hyperbolic Geometry, Quantum Topology and Number Theory. Contemporary Mathematics, vol. 541, pp. 1–40. American Mathematical Society, Providence (2011). doi:10.1090/conm/541/10677

73. Murakami, H., Murakami, J.: The colored Jones polynomials and the simplicial volume of a knot. Acta Math. **186**(1), 85–104 (2001)

74. Murasugi, K.: Jones polynomials and classical conjectures in knot theory. Topology **26**(2), 187–194 (1987). doi:10.1016/0040-9383(87)90058-9

75. Ni, Y.: Knot Floer homology detects fibred knots. Invent. Math. **170**(3), 577–608 (2007). doi:10.1007/s00222-007-0075-9

76. Ozawa, M.: Essential state surfaces for knots and links. J. Aust. Math. Soc. **91**(3), 391–404 (2011)

77. Ozsváth, P., Szabó, Z.: Holomorphic disks and genus bounds. Geom. Topology **8**, 311–334 (2004). doi:10.2140/gt.2004.8.311

78. Ozsváth, P., Szabó, Z.: Link Floer homology and the Thurston norm. J. Am. Math. Soc. **21**(3), 671–709 (2008). doi:10.1090/S0894-0347-08-00586-9

79. Perelman, G.: The entropy formula for the Ricci flow and its geometric applications. arXiv:math.DG/0211159 (2002)

80. Perelman, G.: Ricci flow with surgery on three-manifolds. arXiv:math.DG/0303109 (2003)

81. Petronio, C.: Spherical splitting of 3-orbifolds. Math. Proc. Camb. Phil. Soc. **142**(2), 269–287 (2007). doi:10.1017/S0305004106009807

82. Prasad, G.: Strong rigidity of **Q**-rank 1 lattices. Invent. Math. **21**, 255–286 (1973)

83. Przytycki, J.H.: From Goeritz matrices to quasi-alternating links. In: The Mathematics of Knots. Contributions in Mathematical and Computational Sciences, vol. 1, pp. 257–316. Springer, Heidelberg (2011). doi:10.1007/978-3-642-15637-3_9

84. Reshetikhin, N., Turaev, V.G.: Ribbon graphs and their invariants derived from quantum groups. Comm. Math. Phys. **127**(1), 1–26 (1990)

85. Reshetikhin, N., Turaev, V.G.: Invariants of 3-manifolds via link polynomials and quantum groups. Invent. Math. **103**(3), 547–597 (1991). doi:10.1007/BF01239527

86. Riley, R.: Discrete parabolic representations of link groups. Mathematika **22**(2), 141–150 (1975)

87. Riley, R.: A quadratic parabolic group. Math. Proc. Camb. Phil. Soc. **77**, 281–288 (1975)

88. Scharlemann, M.: Producing reducible 3-manifolds by surgery on a knot. Topology **29**(4), 481–500 (1990). doi:10.1016/0040-9383(90)90017-E

89. Stallings, J.R.: Constructions of fibred knots and links. In: Algebraic and Geometric Topology (Proceedings of Symposia in Pure Mathematics, Stanford University, Stanford, CA, 1976). Part 2, Proceedings of Symposia in Pure Mathematics, vol. XXXII, pp. 55–60. American Mathematical Society, Providence (1978)

90. Stoimenow, A.: Coefficients and non-triviality of the Jones polynomial. J. Reine Angew. Math. **657**, 1–55 (2011)

91. Stoimenow, A.: On the crossing number of semi-adequate links. Forum Math. (in press). doi:10.1515/forum-2011-0121

92. Thistlethwaite, M.B.: On the Kauffman polynomial of an adequate link. Invent. Math. **93**(2), 285–296 (1988). doi:10.1007/BF01394334

93. Thurston, W.P.: Three-dimensional manifolds, Kleinian groups and hyperbolic geometry. Bull. Am. Math. Soc. (N.S.) **6**(3), 357–381 (1982)

94. Thurston, W.P.: A norm for the homology of 3-manifolds. Mem. Am. Math. Soc. **59**(339), i–vi and 99–130 (1986)

95. Turaev, V.G.: A simple proof of the Murasugi and Kauffman theorems on alternating links. Enseign. Math. (2) **33**(3–4), 203–225 (1987)

96. Witten, E.: 2 + 1-dimensional gravity as an exactly soluble system. Nucl. Phys. B **311**(1), 46–78 (1988/1989). doi:10.1016/0550-3213(88)90143-5

97. Witten, E.: Quantum field theory and the Jones polynomial. Comm. Math. Phys. **121**(3), 351–399 (1989)

Index

D. Futer et al., *Guts of Surfaces and the Colored Jones Polynomial*, Lecture Notes
in Mathematics 2069, DOI 10.1007/978-3-642-33302-6,
© Springer-Verlag Berlin Heidelberg 2013

LECTURE NOTES IN MATHEMATICS   🐎 Springer

Edited by J.-M. Morel, B. Teissier; P.K. Maini

Editorial Policy (for the publication of monographs)

1. Lecture Notes aim to report new developments in all areas of mathematics and their applications - quickly, informally and at a high level. Mathematical texts analysing new developments in modelling and numerical simulation are welcome.

   Monograph manuscripts should be reasonably self-contained and rounded off. Thus they may, and often will, present not only results of the author but also related work by other people. They may be based on specialised lecture courses. Furthermore, the manuscripts should provide sufficient motivation, examples and applications. This clearly distinguishes Lecture Notes from journal articles or technical reports which normally are very concise. Articles intended for a journal but too long to be accepted by most journals, usually do not have this "lecture notes" character. For similar reasons it is unusual for doctoral theses to be accepted for the Lecture Notes series, though habilitation theses may be appropriate.

2. Manuscripts should be submitted either online at www.editorialmanager.com/lnm to Springer's mathematics editorial in Heidelberg, or to one of the series editors. In general, manuscripts will be sent out to 2 external referees for evaluation. If a decision cannot yet be reached on the basis of the first 2 reports, further referees may be contacted: The author will be informed of this. A final decision to publish can be made only on the basis of the complete manuscript, however a refereeing process leading to a preliminary decision can be based on a pre-final or incomplete manuscript. The strict minimum amount of material that will be considered should include a detailed outline describing the planned contents of each chapter, a bibliography and several sample chapters.

   Authors should be aware that incomplete or insufficiently close to final manuscripts almost always result in longer refereeing times and nevertheless unclear referees' recommendations, making further refereeing of a final draft necessary.

   Authors should also be aware that parallel submission of their manuscript to another publisher while under consideration for LNM will in general lead to immediate rejection.

3. Manuscripts should in general be submitted in English. Final manuscripts should contain at least 100 pages of mathematical text and should always include

   - a table of contents;
   - an informative introduction, with adequate motivation and perhaps some historical remarks: it should be accessible to a reader not intimately familiar with the topic treated;
   - a subject index: as a rule this is genuinely helpful for the reader.

   For evaluation purposes, manuscripts may be submitted in print or electronic form (print form is still preferred by most referees), in the latter case preferably as pdf- or zipped psfiles. Lecture Notes volumes are, as a rule, printed digitally from the authors' files. To ensure best results, authors are asked to use the LaTeX2e style files available from Springer's web-server at:

   ftp://ftp.springer.de/pub/tex/latex/svmonot1/ (for monographs) and
   ftp://ftp.springer.de/pub/tex/latex/svmultt1/ (for summer schools/tutorials).

Additional technical instructions, if necessary, are available on request from lnm@springer.com.

4. Careful preparation of the manuscripts will help keep production time short besides ensuring satisfactory appearance of the finished book in print and online. After acceptance of the manuscript authors will be asked to prepare the final LaTeX source files and also the corresponding dvi-, pdf- or zipped ps-file. The LaTeX source files are essential for producing the full-text online version of the book (see http://www.springerlink.com/openurl.asp?genre=journal&issn=0075-8434 for the existing online volumes of LNM). The actual production of a Lecture Notes volume takes approximately 12 weeks.

5. Authors receive a total of 50 free copies of their volume, but no royalties. They are entitled to a discount of 33.3 % on the price of Springer books purchased for their personal use, if ordering directly from Springer.

6. Commitment to publish is made by letter of intent rather than by signing a formal contract. Springer-Verlag secures the copyright for each volume. Authors are free to reuse material contained in their LNM volumes in later publications: a brief written (or e-mail) request for formal permission is sufficient.

Addresses:

Professor J.-M. Morel, CMLA,
École Normale Supérieure de Cachan,
61 Avenue du Président Wilson, 94235 Cachan Cedex, France
E-mail: morel@cmla.ens-cachan.fr

Professor B. Teissier, Institut Mathématique de Jussieu,
UMR 7586 du CNRS, Équipe "Géométrie et Dynamique",
175 rue du Chevaleret
75013 Paris, France
E-mail: teissier@math.jussieu.fr

For the "Mathematical Biosciences Subseries" of LNM:

Professor P. K. Maini, Center for Mathematical Biology,
Mathematical Institute, 24-29 St Giles,
Oxford OX1 3LP, UK
E-mail : maini@maths.ox.ac.uk

Springer, Mathematics Editorial, Tiergartenstr. 17,
69121 Heidelberg, Germany,
Tel.: +49 (6221) 4876-8259

Fax: +49 (6221) 4876-8259
E-mail: lnm@springer.com